AIR TRANSPORT PROVISION IN REMOTER REGIONS

Air Transport Provision in Remoter Regions

EDITED BY

GEORGE WILLIAMS
Cranfield University, UK
&
SVEIN BRÅTHEN
Molde University College, Norway

ASHGATE

Published by
Ashgate Publishing Limited
Wey Court East
Union Road
Farnham
Surrey, GU9 7PT
England

Ashgate Publishing Company
Suite 420
101 Cherry Street
Burlington
VT 05401-4405
USA

www.ashgate.com

British Library Cataloguing in Publication Data
Air transport provision in remoter regions.
1. Aeronautics, Commercial--Management. 2. Aeronautics, Commercial--Economic aspects. 3. Aeronautics, Commercial--Environmental aspects. 4. Rural transportation.
I. Williams, George, 1948- II. Bråthen, Svein.
387.7'2-dc22

ISBN: 978-0-7546-7342-2 (hbk)
ISBN: 978-0-7546-9186-0 (ebk)

Library of Congress Cataloging-in-Publication Data
Air transport provision in remoter regions / edited by George Williams and Svein Bråthen.
p. cm.
Papers selected from biennial conferences organized by the Centre for Air Transport in Remoter Regions at Cranfield University.
Includes index.
ISBN 978-0-7546-7342-2 (hardback) 1. Local service airlines. 2. Rural transportation. I. Williams, George. II. Bråthen, Svein.
HE9785A39 2010
387.7--dc22

2010006000

Mixed Sources
Product group from well-managed forests and other controlled sources
www.fsc.org Cert no. SGS-COC-2482
© 1996 Forest Stewardship Council
FSC

Printed and bound in Great Britain by
TJ International Ltd, Padstow, Cornwall

Contents

PART C INFRASTRUCTURE PROVISION

PART D INNOVATION

PART E KEY FUTURE ISSUES

List of Figures

List of Tables

Acknowledgements

We would like to express our sincere thanks to the following who have contributed a chapter (and in some cases more than one chapter) to this book: Göran Anger, David Bentley, Lennart Bergbom, Rodney Fewings, Andy Foster, Knut Fuglum, Nigel Halpern, Johan Holmér, Olav Mosvold Larsen, Jon Inge Lian, Are Lien, Ian Lowden, Basil O'Fee, Romano Pagliari, Rajkumar Pant, Luís Silveira, Jon Sjølander and Pär-Erik Westin.

We also wish to express our gratitude to the following individuals who have played a key role in helping to establish, promote and support the biennial Forum on Remoter Regions Air Transport, from which this book has its origins: Dag Aksnes, Nuno Costa, Knut Fuglum, Keith Hill, Raymon Kaduck, Bob Laird, Bob Macleod, Clarice Rodrigues, Kai Sollid and Ian Stockman.

George Williams and Svein Bråthen

Editors and Contributors

George Williams was until September 2009 Reader in Airline Economics at the Department of Air Transport, Cranfield University. He continues to lecture and research in the fields of Air Transport Economics and Regulation, with a particular focus on the provision of air services in remoter regions. George graduated with a BSc in Economics from City University in 1971, an MA in Transport Economics from Leeds University in 1972 and a PhD from Cranfield University in 1991. He joined the Air Transport Department at Cranfield University as a lecturer in 1995. He has extensive international lecturing experience and has written extensively, exploring the impact of deregulation on the airline industry. In 1998 he was appointed special advisor to the UK House of Lords European Communities Committee enquiry into European Commission (EC) proposals to extend its powers to apply existing competition provisions to air services between the EU and third countries. He has led major studies into airline service provision in Ireland, the Netherlands, Norway, Spain, Sweden and the UK on behalf of the EC, government departments, regulatory authorities, airlines and airport authorities. George heads the Centre for Air Transport in Remoter Regions, which organises a biennial international forum devoted to issues concerning air transport provision in remoter regions, the most recent of which took place in Bergen, Norway, in April 2009. He is a Fellow of both the Royal Aeronautical Society and the Chartered Institute of Logistics and Transport.

Svein Bråthen is a Professor in Transport Economics at Molde University College in Norway, and he also heads the Department of Transport Research at Molde Research Institute. He graduated with a BSc in Civil Engineering (1979), a BSc in Economics (1985) and a PhD in Transport Economics from NTNU, Norway (2001). His research is mainly within Air Transport Economics, Regional Economics, and Economic Impact Assessment. He has led a number of research projects within these areas. Within air transport, Svein has been responsible for developing the methods to be used in Norway for larger air transport investments and he has also undertaken analyses of the Norwegian STOL airport system as well as of the larger airports.

Göran Anger studied Law and Business Administration at the University of Stockholm. His consulting career in Sweden began in the field of Project Management. In 1997 he founded Industrilogik, a consultancy in Logic, Mathematics and Computer Science. Today he is active as an independent consultant in Public Procurement.

David Bentley is a graduate of the Department of Air Transport at the UK's Cranfield University (MSc Air Transport Management) and has been an air transport consultant and writer since 1999. He specialises in airport marketing and financing, having produced several airport route development prospectuses for regional airports, and written six management reports on airport privatisation since 2000. He is also an occasional media commentator, having worked with the BBC, CNN, *New York Times/International Herald Tribune* and Reuters amongst others, and has contributed articles to over 20 newspapers and magazines as well as being the airline correspondent for the *Airfinance Annual*. He is consulting editor of *Airport Investor Monthly*, published by the Centre for Asia Pacific Aviation. Previously he worked in a marketing capacity for three airlines (British Airways, Austrian Airlines and Icelandair); as a partner in a travel agency; and as travel manager for a multinational corporation.

Lennart Bergbom studied Economics at Uppsala University during the 1990s and gained an MPhil in 1998. He then worked with forecasting at the Ministry of Finance in Stockholm before entering the aviation industry at the Swedish Civil Aviation Administration in 2001. He started working with transport policy issues but has gradually shifted his focus towards market analysis and airport strategies. Since 2005 he has been an aviation market analyst at the LFV Group – Swedish Airports and Air Navigation Services.

Rodney Fewings graduated in Civil and Structural Engineering and subsequently gained masters degrees in Engineering Hydrology from Imperial College and Transport Studies from Cranfield University. He worked for many years as a consulting engineer, associated with projects in Africa, the Middle East and the Far East. He joined the Cranfield Air Transport Group in 1987, where he now lectures and undertakes consultancy in air traffic management, airport planning, design and operations and air transport safety. He is a Chartered Engineer, a Fellow of the Royal Institute of Navigation and a Member of the Royal Aeronautical Society and the Chartered Institute of Logistics and Transport.

Andy Foster graduated in Engineering (specialising in Aeronautical Engineering) from Bath University in 1986. He then joined British Aerospace (Bae) Regional Aircraft where he worked in the Technical Sales and Marketing departments, being responsible for producing route performance and cost studies for airlines in Europe, Asia and Australasia. He left BAe in 1994 and then spent two years working on a number of consultancy projects with companies including Touche Ross, SH&E and Air Foyle. In 1996 he joined OAG Worldwide and was responsible for the development and marketing of analytical products based on OAG's airline schedules data – including the CD-ROM based OAG MAX product. Andy joined the Department of Air Transport in 1999 and his academic interests include aircraft performance and economics, airline fleet planning, airline pricing and distribution. In 2003 he became course director for the part-time executive MSc in Air Transport Management and in 2005 for its full-time MSc.

Knut Fuglum graduated in Civil Engineering at Heriot Watt University, Edinburgh in 1970. He worked for some 20 years as a consultant in the Norwegian company AsplanViak, mainly in airport planning, aircraft noise, road planning and public transport. As consultant, he was also engaged in various environmental studies organised by NATO. Knut joined Avinor (the Norwegian airport operator) as senior advisor in 1998, where his main fields of work have been strategic planning, cost-benefit analysis, investment planning, environmental/noise studies, airport/aviation planning and the Norwegian National Transport Plan.

Nigel Halpern is Associate Professor of Transport and Tourism at Molde University College, Norway. He was previously Subject Group Director and Principal Lecturer in Aviation at London Metropolitan University. Nigel joined academia from industry, where he worked for the UK Civil Aviation Authority and Department for Transport, Local Government and the Regions. He has also worked in the tourism and hospitality industries in the UK, France, Spain and Italy. Nigel has an MSc in Tourism Management and a PhD in Air Transport Management. He currently teaches and conducts research and consultancy in transport and tourism, focusing on airports and the development of tourism in remoter regions.

Johan Holmér gained a BSc from Uppsala University and an MPhil from the University of Exeter, focusing on the economic history of air transport regulation and deregulation. He eventually joined the Swedish CAA working with market analysis and air transport policy. After taking a position as manager of market analysis and forecasting at the LFV Group, he worked at the Government Offices as committee secretary and at Rikstrafiken, the Swedish Public Transport Agency, where he was responsible for the tendering of air services subject to public service obligation (PSO). Currently he is senior advisor at Transportstyrelsen, the Swedish Transport Agency.

Olav Mosvold Larsen holds a degree in Political Science from the University in Oslo (UiO). He was previously a researcher at the Program for Research and Documentation for a Sustainable Society (ProSus) at UiO, working mainly on issues related to Environmental Policy Integration (EPI) and sustainable production and consumption. Olav joined Avinor in 2007 as a senior advisor on sustainable development and transport and climate change.

Jon Inge Lian holds an MSc in Geography from Oslo University, 1980. He was four years in Statistics Norway before joining the Institute of Transport Economics, where he has been since 1984, as head of department and now chief research geographer. His main field of interest has been aviation studies (market analysis, forecasts, economic impacts, viability of air routes, effects of airline competition and consequences of deregulation, sustainable aviation). He has also been engaged in regional impacts of infrastructure investments and transport measures in general and has carried out evaluation studies of trunk roads in cities and new fixed links. He has wide international experience, among others from expert groups for the OECD and Euro-Stat and from projects for the EU and the World Bank.

Are Lien was certified as an air traffic controller in 1971. He worked as a controller, supervisor and chief at Oslo ACC until 1989, when he became director of Airport and Air Navigation Services in Avinor. In 1995 he was appointed a member of the Nordic delegation to the International Civil Aviation Organization (ICAO) and worked on several ICAO committees. He then worked as a regional director for six years until he took up the position as project manager, upgrading all Norwegian airports according to new regulations. His field of work has covered most aspects of airport and air traffic management. Are is now working with benchmarking the services at regional airports.

Ian Lowden has an MBA from the University of Brighton with a dissertation on airport business planning, and a first degree in Economic and Social Studies from the University of East Anglia (UEA). He began his aviation career with British Aerospace, forecasting demand for civil aircraft, before working for British Caledonian Airways in route planning. For the last 20 years has been a consultant, providing advice to airlines, airports, investors and regional government with SH&E and then RDG Solutions, part of Route Development Group where he is Managing Director. He has advised many regional and peripheral airports on issues such as forecasting, business planning, non-aeronautical development and the establishment of Route Development Funds, and recently on the impact of Essential Service and Tourism Development Route (ESTDR).

Basil O'Fee was educated in Northern Ireland and was awarded an MA in Geography and Education from Emmanuel College, Cambridge in 1978. He has worked extensively in sales and marketing in the south-east of England, in areas such as gifts, sales promotion and business-to-business multimedia. Moving to Inverness in1996, Basil joined Highland Airways (HA) and helped the company win its first substantial aviation contracts – delivering freight to the Scottish islands transferring and oil crew out of Aberdeen. In 1999 he became a director of Highland Airways and in 2007 was part of the MBO team that acquired the company from the Atlantic Group of aviation companies. Basil was involved in all aspects of the commercial side of HA – its literature, website, public relations strategy, tender responses – and was the driving force of many of Highland's sales and marketing initiatives. Basil managed most of the company's main contracts prior to their subsequent delegation, and remained a major shareholder of the airline until it ceased trading in March 2010. He has been a participant at most of the Aviation in Remote Regions conferences organised by Cranfield University since their inception in 1999. He has presented on the topic of PSOs at two of these conferences.

Romano Pagliari joined the Department of Air Transport at Cranfield University in 1995, having obtained a masters in Transport Studies followed by a PhD in Airport Slot Allocation from the Cranfield School of Management. Romano is currently course director for both full- and part-time MSc programmes in airport planning and management. As an aviation economist, he supports the department's research and consultancy activities. Recent examples include research in the fields of airport infrastructure charging, economic regulation, airport performance benchmarking and air transport in remote regions. Being born and bred in the Scottish Highlands, Romano also has a keen interest in aviation developments in the Highlands and Islands. His other activities include managing and coordinating both public and client-specific airport-related CPD short courses and membership of the scientific committee of the International Centre for Competitive Studies in Aviation and the editorial board of the *Journal of Airport Management*.

Rajkumar S. Pant graduated in Aeronautical Engineering from Panjab Engineering College, Chandigarh in 1983. He then joined Hindustan Aeronautics Ltd (HAL) and was sponsored for a masters in Aeronautical Engineering at the Indian Institute of Technology, Madras. He worked in HAL's Design and Engineering department for five years before joining the Indian Institute of Technology Bombay as a member of faculty in the Aerospace Engineering department in 1989. Subsequently, he earned a PhD in Design, Structures and Systems from Cranfield University in 1997, funded by the Commonwealth Scholarship Commission, UK. His academic and research interests include aircraft design, air transportation and optimisation. Since 2000, he has been involved in developing indigenous technology in India for the design and development of lighter-than-air (LTA) systems for various applications. He is an international member of the LTA Technical Committee of the American Institute of Aeronautics and Astronautics (AIAA), member of the Aeronautical Society of India and a member of the Indian Society for Technical Education.

Luís Silveira was born in the Azores in 1953 and graduated in Law from the University of Coimbra. He subsequently completed several postgraduate studies in Management and several courses in Air Transportation Management. He was a guest lecturer at the University of the Azores Department of Economics and Management and joined SATA, the Azorean Airline, in 1982, where he is now the Commercial Director and an executive member of the Board.

Jon Sjølander graduated in Political Science at the University of Oslo in 1980. He then joined the Norwegian Regional Development Fund and was appointed Avinor's Strategic Director in 1990. He has been a member of the company's senior management for almost 20 years, holding positions such as Director of Planning and Development and Director for Avinor's Technical-Operational Division. His main fields of work have been strategic planning, air traffic management, environmental questions, investment planning, decision-making and reorganisation. He has delivered numerous lectures on aviation both in Norway and abroad. Jon has also been the Chairman of Oslo Airport's Board of Directors.

Pär-Erik Westin has BSc in Business Administration and Economics. Having gained initial experience with audit firm Ernst & Young he worked on the liberalisation of the telecommunications sector with the national Swedish regulator for seven years. Following this he migrated to another infrastructure-based industry, namely public transportation. He was instrumental in designing the Swedish model for the tendering of air services subject to public service obligations, and he has hands-on experience of all phases of tendering. Currently he is on leave from the national agency Rikstrafiken to work for the Government Offices as an expert in the field of public transportation.

Introduction

George Williams

The origins of this book lie in a study that the Air Transport Group at Cranfield University undertook back in 1997/98 on behalf of and in collaboration with the then Norwegian Civil Aviation Administration (Luftfartsverket). The purpose of the study was to compare the significance of air transport and its provision in terms of airport infrastructure in a selection of European countries, to investigate how different airport systems are organised and managed, to compare the cost of providing airport infrastructure and related services, and to examine the importance of aviation as a means of public transport for domestic journeys.

In all, 429 airports situated in 13 countries were examined in the study. While a great deal of data was collected and analysed, it was clear that considerably more work remained to be undertaken before a detailed and accurate comparison could be made of the financial performances of different airport systems. One conclusion of the study was that there was a clear need for further research examining the relative efficiencies of different forms of airport systems. In addition to this, the need for further research on market conditions, air transport policies for remoter regions, aircraft operations and network design and operations became evident. It was undoubtedly the case that lessons could be learnt from the different approaches to air transport provision across Europe. A view emerged that a forum for the exchange of such information would provide an effective means by which to continue the work undertaken in the study. By obtaining the interest and experience of a wide range of airport system operators and organisers, much progress could be made to ensure the dissemination of best practice and the efficient use of scarce resources.

The first Forum on Air Transport in Remoter Regions was held in Nairn in Scotland in 1999, with the aim of bringing together those responsible for managing, operating and regulating both airports and airline services in the remoter parts of Europe. The programme included speakers from Finland, Greece, Iceland, Ireland, Norway, Portugal, Scotland, Spain and Sweden. Their wealth of experience and knowledge gave a unique opportunity for extensive discussion and exchange of ideas on an often-neglected area of air transport.

The first two days of the Nairn Forum were devoted to issues concerning Europe's smaller regional airports, namely: their organisation and operation, their planning and their economic and social impact. Presentations were given by those responsible for managing and operating these types of airports in Finland, Greece, Iceland, Norway, Scotland and Sweden. The final day began by focusing on the operation of air services in remoter regions and closed with an examination of the different forms of subsidisation that have been adopted.

Following the success of the first forum the event has become a biennial event with a wider coverage than simply Europe. The most recent forum was held in April 2009 in Bergen, Norway. As with previous meetings, representatives of airlines, airport authorities, central and local government departments and agencies, consultancies and academic institutions from across the world were in attendance. On this occasion 20 speakers spoke on five key topics affecting the sector, namely: Environment and Sustainability, Subvention Policy, Safety and Crisis Management, Airport Infrastructure, and Innovation and Managing Change.

The sharing of ideas and experiences of best practice in the provision of air transport from the more remote parts of the world – including Brazil's Amazon region, the Azores, the Faeroes, Lofoten and Finnmark in Norway and northern Canada – has without doubt proved of benefit to those charged with operating airline services and providing the necessary infrastructure in regions where demand for air transport is small but vital for the well-being of the communities living there.

The purpose of this book therefore is to provide the reader with a flavour of some of the many topics of concern to those responsible for operating, managing, financing and planning the provision of air transport in remoter regions that have formed the subject matter of these biennial forum meetings. As with each forum, the book focuses on five key topic areas, namely: the provision of air services and their impact (Part A), subvention mechanisms (Part B), infrastructure provision (Part C), innovation (Part D) and key issues for the future (Part E). In all, some 20 experts, each with extensive knowledge of operating, regulating, planning or funding air transport in regions with low demand for air transport, have expressed their thoughts and opinions.

In Part A, Svein Bråthen from Molde University College examines the key theoretical issues involved in deciding how much air transport should be supplied in remoter areas; Romano Pagliari from Cranfield University and Luís Silveira, the commercial director of SATA Air Açores, explore the development of air services over the past two decades in, respectively, the Scottish Highlands and Islands and the Azores; Jon Inge Lian from Norway's Institute of Transport Economics analyses the economic impact of air transport in remoter regions; and Nigel Halpern from Molde University College reviews the marketing of small regional airports.

In Part B, George Williams from Cranfield University examines the experience across Europe of the application of Public Service Obligations (PSOs); Göran Anger, Johan Holmér and Pär-Erik Westin from Sweden's National Public Transport Agency (Rikstrafiken) explore in detail the tendering process experience of PSOs in Sweden; and Basil O'Fee, the commercial director of Highland Airways, describes his experience of tendering for and operating PSO routes.

In Part C, Rodney Fewings from Cranfield University reviews the provision of airport infrastructure in Europe's remoter regions; Are Lien from Avinor analyses the problems for smaller regional airports of complying with International Civil Aviation Organization (ICAO) Annex 14; and Svein Bråthen from Molde

University College and Knut Fuglum from Avinor explore the issues involved in developing an airport located in northern Norway over the longer term.

In Part D, Lennart Bergbom from Luftfartsverket and Johan Holmér review the performance of the virtual airlines that have been established in Sweden; Rajkumar Pant from the Indian Institute of Technology Bombay analyses the economic case for using airships to transport passengers and freight in remote parts of India; and consultants Ian Lowden and David Bentley provide an overview of the ICAO's development scheme for routes with tourism potential.

In Part E, Olav Mosvold Larsen and Jon Sjølander from Avinor examine the issues affecting aviation providers faced with the challenge of climate change; Andy Foster from Cranfield University reviews the options available for airlines seeking to replace their ageing sub-20 seat aircraft; while George Williams explores the impact of measures that have been adopted in Scotland to reduce the high airfares charged to passengers travelling to the remoter parts of the country and Svein Bråthen presents a study from Norway which indicates that airfare reductions on thin routes may have socio-economic benefits for Norway as well as favourable effects for the remoter regions.

That there are many challenges facing those responsible for providing air transport services in the more remote parts of the world is obvious, and this book will hopefully shed light on a part of the transport sector which provides lifeline services for many small communities. A great number of these communities play active roles in their national economies by providing manufacturing skills (e.g. the oil and gas industry), services (e.g. tourism) and natural resources (petroleum and fish). Such issues will probably become more important also in emerging economies such as China and India.

This book is dedicated to the ingenuity, determination, resourcefulness and sheer endurance of the large band of professionals who master these difficulties on a daily basis in conditions that many who live in larger communities would find simply impossible.

Chapter 1

Deciding Upon the Right Amount of Air Transport Services in Remoter Regions – Key Issues

Svein Bråthen

Introduction

Air transport services are a key factor in the transport system in rural areas. They allow human and natural resources to be used efficiently in a modern society. Air transport is very important for long-distance travel and for connecting remote areas to the rest of the world. In order to provide air transport, the air transport industry utilises primary resources, such as labour and capital, as well as intermediate products. Air transport itself is a product which is delivered partly as an intermediate product for other industries and partly for final demand like leisure travel.

The air transport industry, as well as other production activities, has impacts for resource allocation as well as for value added and income. Air transport services also act as a catalyst in industrial development, particularly in connection with the exploitation of location-specific natural and human resources. These catalytic impacts could be very important both from an economic efficiency and regional balance perspective. They are further discussed in Chapter 4.

To be able to provide efficient air transport services, it is important to have an adequate balance between level of service, air fares and subsidies. Excessively high prices may cause undesirable traffic deterrence and hence the benefits of the resources that are put into the air transport services are not fully utilised. On the other hand, if fares are set too low and/or the level of service is too high, then this may give excess demand and/or too many resources may be put into the air transport system.

In this section, we will discuss some key issues with respect to how the level of service in air transport can be determined, and what the economic consequences of a well-functioning air transport service are. The link between costs, prices and economic efficiency will be an important part in this discussion. Theoretically, the trade-off between fares, subsidies and departure frequency should be balanced so that the marginal willingness to pay for the air services should be equal to the costs of the marginal waiting time for the passengers. Firstly, market imperfections, costs issues and the balance between fares, costs, level of service and subsidies will be discussed. Secondly, tendering as a means of efficient provision of airport services will be briefly presented. Finally, air transport services and the impact on economic efficiency in remoter areas will be commented on.

Market Imperfections and Economic Efficiency

Air transport causes a lot of effects which may call for some kind of attention by the public authorities. One of the analytical approaches that is appropriate when dealing with various problems in air transport economics is applied microeconomics.

In general, the authorities are involved in airline markets because they may fail to achieve economic efficiency, mainly with respect to these three categories:

- inability to provide public goods in line with the society's preferences;
- the presence of externalities;
- imperfect competition.

Air transport services can be seen as a kind of public good as long as there is excess capacity, and society may have preferences for offering them in remoter areas for equity reasons. External effects are elements of costs and benefits that are not or only partly priced in the market. Examples may be costs of CO_2 emissions and benefits from improved communication in, for example, industrial networks. In most cases, imperfect competition may cause air fares to rise well beyond marginal costs. Imperfect competition based on natural monopolies (increasing returns to scale) may also call for subsidies to maintain economic efficiency.

In this section, we are going to focus mostly on the third of these elements. Indirectly, this element is also a discussion of the right amount of services to be provided in remoter areas; hence the first element will also be discussed. The main source of inefficiency is when the price of a product or service is set differently from the marginal social costs. On low-volume routes, the difference between price and marginal cost may become high, probably too high, without some kind of money transfer, either by means of cross-subsidies or by public subsidies.

Prices and Subsidies in Air Transport

We are now going to discuss theoretically what the demand and supply side of air transport in remoter areas may look like. In Figure 1.1, price P_m with the corresponding quantity Q_m is the situation for a profit-maximising monopolist, such as an airline that has a licence to operate a given Public Service Obligation (PSO) route.

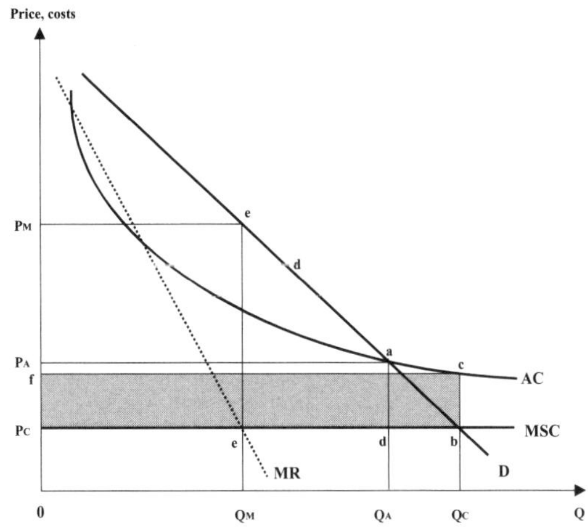

Figure 1.1 Output and pricing policies in a natural monopoly

The production takes place as long as the marginal revenues (*MR*) exceed the marginal costs (*MSC*) at point e. If the monopolist is *regulated* by the authorities to e.g. charge a price equal to the average cost (*AC*), then the price will be P_A and the output will be Q_A. However, the social optimum is where the price equals the marginal cost (*MSC*), with the corresponding output Q_C. The fixed costs are however not covered, and the financial *loss* will be the shaded area $P_e fcb$. This loss has to be covered by subsidies. A *first-best* pricing regime (price equals the marginal cost) gives a financial loss in a situation with a natural monopoly (increasing returns to scale). A *regulatory* regime where the monopolist is allowed to charge the price P_A, incurs a socioeconomic loss equal to the area *abd*. It is worth noting that in many cases, the capacity is well in excess of demand, meaning that the intersection between the *D* curve and the *AC* curve in Figure 1.1 takes place where the *AC* curve is well above the *MSC* curve. In those cases, a pure *AC* pricing policy will incur a higher P_A, a significantly larger *adb* triangle and hence the economic welfare loss to society will also be larger. The slope of the demand curve is also of interest. A steeper, less price-elastic curve will cause a smaller welfare loss.

Price Elasticity Is An Important Measure

The price elasticity gives percentage change in demand following percentage change in prices, and is used for assessing, e.g., revenue effects of air fare adjustments. As seen below, the price elasticity is an important measure when assessing efficient prices and level of service in air transport in remoter areas.

There are not many studies of price elasticity for air transportation. The most recent one is Njegovan (2006), who cites a number of other studies. One of them is Dargay and Hanly (2001) who used a time series from 1989 to 1998 to estimate a price elasticity of -0.6 for outbound international tourism from the UK. An interesting feature of this study was that other factors like currency parity and relative prices affected the demand for air services even more. Njegovan (2006) found the price elasticity in the same market to be -0.7, based on a time series from 1993 to 2003.

One of the most cited studies is Gillen et al. (2003), who examined 21 studies of price elasticity in various air transport segments in different countries. These studies were mainly based on time series analyses. This study is also the main source of information to support European Commission (2005), where the consequences of fuel taxes in the European Economic Area are discussed. Figure 1.2 summarises the findings from Gillen et al. (2003).

The variations appear to be quite large. Short-haul domestic (which is a significant part of the market for air transport services in remoter areas) has a median of -0.7, varying between -0.6 and -0.8. Domestic leisure travel is considerably more elastic, with a median of -1.5, varying between -1.3 and -1.7. A Norwegian study (Helgheim, 2002) of two domestic markets shows a price elasticity of mainly between -0.6 and -1.0, with a leisure travel segment ranging as high as -1.4. The interesting feature of this study was that it indicated an increasing elasticity with higher air fares, which is in good accordance with economic theory. Rekdal (2006) used network model simulations indicating somewhat lower price elasticity for air transport in the range of -0.41 to -0.76.

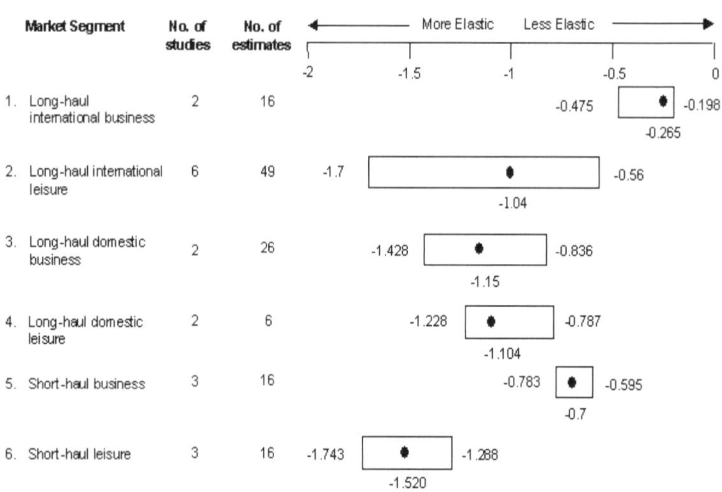

Figure 1.2 Variations in direct price elasticity for air transport services

Source: Gillen et al. (2003).

Optimal Pricing of Air Transport Services with Subsidies

In connection with regional air services, the trade-off between economic efficiency (price P_c and the economic loss $P_C fcb$) and cost coverage (price P_A with traffic loss Q_C-Q_A) will often end in a discussion of compromising between public financial outlays and making the air transport services reasonably priced to make the system serve the remoter regions in an adequate way. This discussion tends to be more important on routes where the payload is low and where a minimum level of service has to be provided (less demand causes the intersection between the average cost curve *AC* and demand *D* to be well towards the left of point *a* in Figure 1.1), as compared to the PSO routes where traffic is rather high (intersection between *AC* and *D*, say, between point *c* and *a* in Figure 1.1). One way to do this is to price the services optimally, however constrained by the maximum level of subsidies the public authorities are willing to pay. In practice, this means that the authorities are willing to cover some of the shaded area in Figure 1.1, but not all of it. We call this pricing strategy the optimal departure from marginal cost pricing, or second-best pricing. We talk about a *financial constraint*, namely the maximum subsidy that the authorities are willing to accept or, conversely, the minimum revenues that have to be collected from the passengers to ensure that the 'maximum subsidy' criterion is met and at the same time maximise economic efficiency. We are now interested in how this can be done so as to minimise the economic welfare loss (the 'deadweight loss').

Consider a 'thin' regional air transport route where there are two market segments in a monopolised market. A reasonable assumption is that these markets have different price elasticity. The demand $x_i = x_i(p_i)$ is dependent on the price level. Conversely, the prices are dependent on the demand:

$$(1) \quad p_i = p_i(x_i)(i=1,2)$$

The cost function for both market segments is expressed as:

$$(2) \quad C = c_1(x_1) + c_2(x_2) + B$$

C is total costs, B is fixed costs, $c_i x_i$ are variable costs dependent on traffic for each market segment.

The financial constraint can now be expressed in a way that the revenue minus fixed and variable costs should at least equal the revenue necessary to satisfy the maximum subsidy constraint:

$$(3) \quad [p_1 x_1 + p_2 x_2] - [c_1(x_1) + c_2(x_2) + B] \geq \Pi$$

We are now going to maximise the consumer surplus (the area between the demand curves and the cost curve), given the budget constraint (or subsidy constraint) Π. This can be expressed as:

$$\text{(4) Maximise} \quad \int_0^{x_1} p_1(x_1)dx + \int_0^{x_2} p_2(x_2)dx - c_1(x_1) - c_2(x_2) - B$$

given the constraint

$$p_1x_1 + p_2x_2 - c_1(x_1) - c_2(x_2) - B \geq \Pi$$

By using Lagrange's method for partial derivation, we get:

$$\text{(5)} \quad \frac{p_1 - c_1'}{p_1} . e_1 = \frac{p_2 - c_2'}{p_2} . e_2$$

Equation (5) is useful in practical pricing policies. p_1 and p_2 are the air fares in each market segment, e_1 and e_2 are the price elasticity for each segment and c'_1 and c'_2 are the marginal costs. The economic interpretation of the equation is that if there are binding financial constraints, prices should be set higher than the marginal costs. The difference from marginal cost pricing should be higher in the market segment with the lowest price elasticity, implying that the price is set higher in the segment with the least price sensitivity. This is done to ensure a minimum welfare loss. The achievement is therefore important because the welfare loss for the consumers in each market segment is minimised.

The price elasticity in each segment has to be different in order to differentiate the prices, but the marginal cost can be equal (which is normally the case in an aircraft with two market segments). Imagine that market 1 is the market for business travel (low price elasticity) and market 2 is the market for leisure travel. The marginal costs and the demand elasticity are normally known, and the point of departure is to set the prices in one of the segments as the 'target price'. For example, one might want to have a low fare for leisure travel. Equation (5) can then be used to calculate the price in the other market segment in order to satisfy the subsidy constraint. The subsidy constraint can be determined by considering the average variable costs per passenger that are caused by each departure. The difference between total costs and the average variable costs per departure can be considered as the optimal subsidy level, given that the correct level of service (LOS) is established, such as aircraft size and departure frequency. The correct LOS can be determined by studying the level where the waiting time costs for the passengers is equal to the costs of an incremental increase in the number of departures. By doing these considerations, optimal fares can be set, dependent on the capacity offered in the market, the price elasticity and the marginal costs of providing the service.

There are of course elements that make this a bit more complicated. One is the level of service that should be provided and the corresponding aircraft costs. In general, pricing policies should aim at maximising the payload for each departure. In general, empty seats represent a loss to society. Either the capacity is above optimality or the air fares are set too high. The pricing framework described above offers a reasonable compromise between marginal cost pricing and the need to contain public spending on air transport services.

More often than not, determination of LOS is based on fairly simple, intuitive assessment because of limited information about the market segments and their ability to carry the economic costs of the services. The subsidy level may also be set based on equity or regional balance considerations and not only from criteria based on economic efficiency. However, cost-benefit analyses can be applied to assess the adequate level of air transport services, based on the criteria above.

A study by Jansson (2007) indicated clearly that a 20 per cent reduction in air fares for the PSO routes in Norway would give a better overall economic efficiency because of the passenger benefits and positive system effects in the transport network in general. In addition to the overall net efficiency gains, the regional balance effects may be considered as positive given a political objective to favour remoter regions: passengers in the PSO system are expected to be the winners. These results underline the importance of getting the prices right in remote air transport services. However, this study did not aim at optimising the level of service.

Aircraft Operating Costs

Aircraft operating costs are one of the most important factors when determining the LOS. The cost structure of an airline can be considered in different time-scales such as day, week, month, year, which is dependent on the purpose. It can be shown that airline costs have a stepwise upward sloping curve, where the steps are caused by aircraft capacity expansion.

As a practical approximation to these costs, Janic (1999) has estimated a regression model to quantify the average costs per flight dependent on the aircraft size/capacity and non-stop route length as follows:

$$C(n,d) = 7.934 \cdot n^{0.603} \cdot d^{0.656}$$

where *C(n,d)* is average costs per flight;
n is aircraft seating capacity; and
d is route length.

As can be seen, the model indicates the existence of economies of scale both with respect to route length and aircraft size. Data from 21 western European airlines are used, and the model explains nearly 90 per cent of the variations. The statistical diagnosis suggests that the equation and the coefficients are significant at the level of 5 per cent and 1 per cent. The coefficients, which have values less than 1, indicate that there are economies of scale both with respect to aircraft size and route length. Figure 1.3 shows the economies of scale with respect to route length for different aircraft, with flight distances up to 400 km. The figure also indicates the scale effects related to aircraft size.

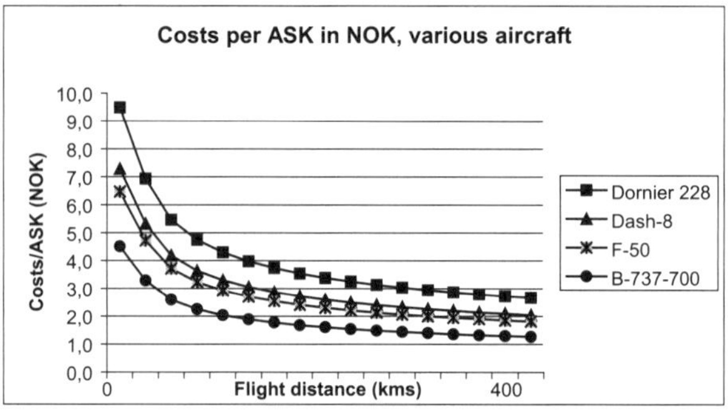

Figure 1.3 Aircraft costs per available seat kilometre (ASK) related to aircraft size

This cost model is estimated for aircraft size of around 100 passengers and above, and at the outset it is not applicable for smaller aircraft. To test this, the model (with an adjusted constant) was compared to the Norwegian cost data from the winners of the competitive tendering process in the year 2000. The comparison comprised 19 of the PSO routes. The model results deviated by only 2 per cent for the 19 routes in total, in the sense that the costs were slightly overestimated in the model. The model results were mainly within ±20 per cent at the route level, compared to the reported route costs from the competitive tendering process.

The properties of the model seem appealing in the sense that it incorporates the scale effects of flight length and aircraft size, and because it gives a reasonable fit to the data for actual route costs. The model has been used in the calculations of changes in aircraft operating costs as a reasonable approach when assessing changes in the airport network.

Competitive Tendering as a Means of Controlling Airline Costs

In Chapter 6, the Public Service Obligations mechanism as a means for offering subsidised air transport routes is described in more detail. Economic theory states that under given assumptions (free entry and exit, prices are known and correct, atomistic price takers etc.), competitive markets will ensure efficient resource allocation. If one or more of these assumptions are violated, imperfect markets and market failure will result. Under such circumstances there may be cases for public sector intervention. A classic example is when markets are characterised by increasing returns to scale (IRS), where competition in the market will cause inefficient duplication of fixed costs. Air transport services in remoter areas have an IRS cost structure where in most cases there is room for only one airline to serve the routes. Hence, some kind of public concession operation seems adequate to reduce the fixed costs. On the other hand, the operator gets significant market power which calls for some kind of air fare regulation.

Public regulation of companies is not a straightforward matter because there is always a possibility of asymmetric information between the agent (the airline) and the principal (the authorities). As a consequence, Demsetz (1968) proposed that in cases where market imperfection does not allow for efficient competition, competition *in* the market should be replaced by competition about *access* to the market. A tendering process is a competition about a concession to serve a specific market.

In the tendering processes, the dimensioning of the air transport services (departure frequency, maximum air fares and aircraft capacity) has been made by the authorities. Ideally, the dimensioning of the air transport services should be made by means of the theoretical points described above. How these tenders will work out in the long run in terms of enhancing sound competition about these concessions will, to a certain extent, depend on the number of players and also on how routes may be bundled in the tender processes, if there is a route network where such bundling is feasible. If the links are tendered separately, there are reasons to expect a larger number of bids than if several links are bundled in one tender. Larger bundles subjected to tendering will probably exclude smaller companies from entering. Relatively few players in repeated tendering processes may cause the improvements in internal efficiency to be smaller than what can be observed in the initial rounds of tendering, even if the routes are announced in separate tenders. It is well known – see e.g. Thredbo 7 (2002) – that imperfect competition through merging and acquisition is a major threat against tendering and deregulation as means for reducing costs. In some of the markets for air transport services in remoter areas, the lack of competition in the tendering processes has been a matter of concern.

In markets where a limited number of competitors meet each other in repeated tendering processes, collusion behaviour might occur. The incentives for collusion between the competing carriers is mainly caused by the danger of losing significant profits if one of the players reveals the true costs by undercutting the profit-maximising price (Demsetz, 1968; Church and Ware, 2000; Thredbo 7, 2002). There is a paradox that the incentive to collude may be strengthened if a larger number of players are involved. The reason for this is that more aggressive price-cutting behaviour can become the result with more players involved, particularly if asset specificity occurs. Arguably, the air transport market may not be characterised by a high degree of asset specificity because aircraft may be used elsewhere if bids are lost; albeit there are some fixed costs involved in setting up the route and gaining the necessary market knowledge. However, incentives to collude are particularly strong where economies of scale and/or excess production capacity exist. Elements of this cost structure are present in air transport services. In the longer run, the relatively small number of players in many markets, combined with a prospective significant level of asset specificity, may spur some kind of collusion through evolutionary processes in repeated games.

A foreign entrant will need to invest a significant amount in factors such as administration, market studies, aircraft, crew and information systems. These transaction costs make it likely that foreign players will enter with the ambition of being profit-maximising going concerns. Hence, foreign companies that want to participate in national tenders can also be expected to participate in a prospective collusive game in the longer run. Nevertheless, it will be important for the authorities to have a critical view on the barriers to entry to facilitate new entries. It is also worth noting that even if collusion behaviour can be expected from sound theoretical reasoning, these authors know of no empirical evidence on collusive behaviour in the air transport market.

Airport Costs

Airport charges are supposed to cover the costs of providing the airside part of the airport services. These charges are included in the air fares, and hence they are a part of the discussion about efficient air fares. These charges are normally set from an average cost perspective and the non-airside revenues may also affect the level of airside airport charges. The average costs may be calculated from a total cost covering perspective for the airport network as a whole, like in Norway. Hence, the airport charges only rarely reflect the marginal costs of an aircraft movement, which should ideally be the efficient level of the charges.

Airports in remoter areas normally have no capacity problems; hence the cost structure gives increasing returns to scale. In such systems, the average cost of providing airside services falls until the capacity limit is approached. This means that once the fixed costs are taken to operate the airports (infrastructure, maintenance and personnel), the marginal costs (*MSC* in Figure 1.1) are mainly constant up to the capacity limit, and they are rather low. As seen in Figure 1.1, charges according to average costs may then give a socio-economic loss to the society equal to the area *abd*. If the demand *D* is smaller than shown in the figure, then the area *abd* may become much larger.

Charges equal to average costs may cause traffic deterrence and hence less traffic than optimal. In 2002, the Norwegian Competition Authority (NCA) made a formal enquiry to Avinor's predecessor, Norwegian Air Transport and Airport Management, about the charging structure of Norwegian airports in remoter areas. The NCA claimed that the charges should to a larger extent reflect the marginal costs of the airport services to avoid traffic deterrence and to improve the use of the airport and aircraft capacity. The NCA argues that a pricing system based on optimal pricing under financial constraints (as described above) should be applied. The NCA also argued that the Norwegian Ministry of Transport and Communications should consider increasing the airport subsidies on airports with thin routes to avoid excessive airport charges.

Air Transport and Regional Development

We have now discussed a few issues connected to correct pricing of air transport services in remoter areas, both at the airline and airport level. Air transport services are of great concern to local communities because there is a strong belief in air transport services as an important prerequisite for regional development. Figure 1.4 illustrates some of the reasons why air transport services are considered important. The underlying case is that accessibility to markets for supplies and finished goods and services is radically improved for many remote areas such as, e.g., the Highlands and Islands of Scotland and northern Norway. Accessibility for private households in remoter areas will also most likely be important for being able to maintain welfare, e.g. by having the possibility to visit friends and relatives and to be able to participate in the social life also in more urban areas (Båtevik et al., 2002, Bråthen, 2003).

Figure 1.4 shows possible causes and effects if an air transport service improves the conditions for local industry. Improved accessibility may make firms relocate, which in turn augments the demand for labour and capital.

In Figure 1.4, there are four main loops that describe the circular and cumulative effects in principle. Firstly, in-migration of firms increases the population and hence the local supply of skilled labour. Secondly, the increase in local employment and population makes capital and firms move to the area to serve the demand for goods and services. Thirdly, secondary industries may establish themselves or increase

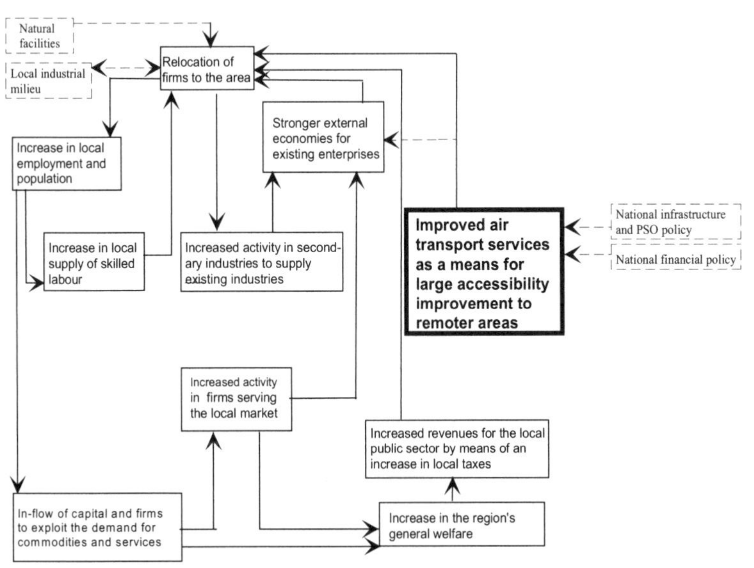

Figure 1.4 Air transport services/improved accessibility and regional development

their activities because of a growing market. Both the second and the third loop may support the development of external economies for existing enterprises. The external economies evolve from a larger number of interacting economic agents. The dotted line from infrastructure to external economies indicates the infrastructure's role in reducing 'the costs of interaction' between agents. The fourth loop illustrates how the larger economic activity augments the public sector revenues, which again may make location in the area more favourable. However, one of the main unsolved empirical issues in general is related to the 'critical mass problem', i.e. what level of economic activity has to be present to maintain viable local communities.

References

Bråthen, S. (2003). Luftfartens rolle i regional samferdsel. Avinor/Molde Research Institute.

Båtevik, F.O., G.M. Olsen and B. Vartdal (2002). Det gode liv – for folk Vest. Ein studie blant nøkkelpersonar på arbeidsmarknaden i Møre og Romsdal. Volda Research Institute, Working Paper 7/2002.

Church, J. and R. Ware (2000). *Industrial Organization: A Strategic Approach.* Boston, MA: McGraw-Hill.

Dargay, J. and M. Hanly (2001). The Determinants of the Demand for International Air Travel to and from the UK. Paper presented at the 9th World Conference on Transport Research, Seoul.

Demsetz, H. (1968). Why regulate utilities? *Journal of Law and Economics*, 11/1, 55–65.

European Commission (2005). New Sources of Financing for Development: A Review of Options. Commission Staff Working Paper SEC, 467, Brussels.

Gillen, D.W., W.G. Morrison and C. Stewart (2003). Air Travel Demand Elasticities: Concepts, Issues and Measurement. Department of Finance, Government of Canada.

Helgheim, B.I. (2002). Norsk luftfart. Avgifter og elastisiteter. MSc thesis in Logistics, Molde University College.

Janic, M. (1999). Behaviour of Western European Scheduled Airlines during the Market Liberalisation Process. In M. Beuthe and P. Nijkamp (eds), *New Contributions to Transportation Analysis in Europe.* Aldershot: Ashgate Publishing, pp. 317–42.

Jansson, K. (2007). Konsekvenser av større andel statlig finansiering av luftfarten i Norge. Molde Research Institute, Report 0714.

Njegovan, N. (2006). Elasticities of demand for leisure air travel: A system modelling approach. *Journal of Air Transport Management*, 12, 33–9.

Rekdal, J. (2006). Evaluation of the Norwegian long distance transport model (NTM5). Molde Research Institute, Report 0609.

Thredbo 7 (2002). International Conference on Competition and Ownership in Land Passenger Transport: Summary Reports. *Transport Reviews*, 22/3, 335–70.

Chapter 2

Trends in Air Service Development within the Highlands and Islands of Scotland 1983–2006

Romano Pagliari

Introduction

The economic and social development of peripheral regions is highly dependent on the provision of adequate and easily affordable transportation services. In some regions of the world, because of geographic location, climate and physical terrain, air transport is the only viable means of transportation for both goods and people.

Peripheral regions have two key features: low population densities and a lack of proximity to major economic and political centres. Motivated primarily by the political necessity to maintain cohesion within national jurisdictions, special arrangements are usually made with regard to the provision of air services to and within these regions. The high costs associated with providing these services, coupled with the need to maintain and operate airport infrastructure to serve small communities, means that subsidy, in the form of grants from national, regional and local levels of government, is normally required to sustain a satisfactory level of air service provision.

Methods of paying subsidy and the sources of funding vary by country. For example, in the United States there exists a comprehensive system of subsidies organised within an Essential Air Services Programme. This was established in response to concerns regarding the future viability of air services to small communities as a result of deregulation within the domestic airline market. Research undertaken during the late 1980s and early 1990s appears to validate these concerns.

Policy-makers in the European Commission (EC), in formulating legislation to liberalise air services, implicitly recognised the possible risks to the future viability of air services to peripheral communities with provisions for the imposition of public service obligations (PSOs) (Commission of the European Communities, 1992). These provisions grant national governments of member states the authority to protect and subsidise routes that are of vital socio-economic importance. Levels of subsidy and decisions on which routes qualify for PSO status are left to the discretion of each member state. According to Reynolds-Feighan (1995), it is because of this that significant regional distortions exist in the quality and availability of air services to and within Europe's remote regions. For example, in France approximately 80 domestic routes have PSOs imposed on them. In Scotland, on the other hand, PSOs are imposed on a limited number of ultra-peripheral routes. Variations in the degree to which member states have applied PSOs have been highlighted by Williams and Pagliari (2004), in particular the comparatively low number of PSOs in Scotland relative to France.

Public service obligations, supported by subsidy, can guarantee for a fixed period of time continuity of air services and low fares on routes linking peripheral regions with hub airports. In the absence of PSOs, these markets are subject to shifting airline strategic priorities in response to developments in global markets. Such a shift in priorities can be manifested in increased management focus on providing services on higher density profitable networks while simultaneously rationalising service levels and delivery on costly and generally lower margin regional air services. Rationalisation can be achieved through franchising, where major airlines sub-contract unprofitable segments of their networks to dedicated regional carriers. However, franchising also allows an airline to retain the presence of its brand in a franchised market (Denton and Denis, 2000). Air services that are of vital socio-economic importance to communities located in peripheral regions, without regulatory protection, are therefore vulnerable to changes in levels of provision as a consequence of these developments.

Between 1994 and 2000, as a result of sustained losses incurred from directly operating air services in the Highlands and Islands of Scotland, British Airways (BA) initiated a process of franchising its Highlands and Islands operations to regional carriers. Initially services were transferred to two airlines: Manx and Loganair. Gradually over time, Loganair was to assume responsibility for all British Airways services in the Highlands and Islands of Scotland. It is interesting therefore, to ascertain to what extent franchising has affected the level of air service provision in the Highlands and Islands.

Air Transport in the Highlands and Islands

Socio-economic Environment

The Scottish Highlands and Islands account for 13 per cent of the land area of the British Isles. A sizeable proportion of this area consists of islands and mainland communities located at a significant distance from key political and economic decision-making centres in both the United Kingdom and wider European Union. The population of the Highlands and Islands, according to the latest statistics, is 440,761, with 61 per cent living in settlements of fewer than 5,000 inhabitants (Highlands and Islands Enterprise, 2002). Table 2.1 provides population statistics for a number of Scottish islands between 1981 and 2001. It is apparent from these statistics that all the island communities listed, with the exception of Orkney, experienced a decline in population between 1981 and 2001. For example, Islay's population has declined by 14 per cent over the entire period, while on the larger island of Lewis and Harris, the population has fallen by 8 per cent. Worryingly, from an economic development perspective, the General Register of Scotland predicts a 13 per cent decline in the population of the Western Isles between 2000 and 2016 (Western Isles, 2002).

The Highlands and Islands region is highly dependent on air travel where climatic conditions and topography can make direct surface routes tortuous and inconvenient, where sea fogs, high winds and heavy precipitation impact on the punctuality and regularity of scheduled transportation services (UK Civil Aviation Authority, 1974). Table 2.2 provides a comparison of air service journey times with surface modes of public transportation between selected locations. For example, a flight time of one hour between Glasgow and Stornoway in the Western Isles compares to a surface journey time on public transport services of just over eight hours involving three different modes: ferry, coach and rail. This serves to illustrate the significant time advantage air services enjoy over other modes and how dependent these regions are on air services to ensure convenient access to mainland population centres.

Due to low population density (9 persons per square kilometre compared with an EU average of 116), economic underdevelopment and the relative inaccessibility of communities situated in the Highlands and Islands, commercial air transport in this region is at worst significantly unprofitable and at best marginally economic. This has been the situation ever since the first scheduled passenger services were launched in the 1930s. According to a report on air services in the Highlands and Islands of Scotland (UK Civil Aviation Authority, 1974), three out of 18 services operated by the then Scottish Airways generated an operating profit (Glasgow–Benbecula, Aberdeen–Wick and Inverness–Wick). Nearly 30 years later, despite attempts during the early 1980s to rationalise services, according to Kyrou (1998), British Airways was losing approximately US$5 million per annum on directly operating air services between the Scottish mainland and the Northern and Western Isles.

Table 2.1 Population statistics: sample airport catchment areas 1981–2001

Regions	Airport	1981	1991	2001	% change 1981–2001
Lewis & Harris	Stornoway	22,485	21,737	19,918	-11.4%
North & South Uist	Benbecula	3,630	3,510	3,138	-13.5%
Orkney	Kirkwall	18,419	19,612	19,245	4.5%
Shetland	Sumburgh	22,768	22,522	21,988	-3.4%
Islay	Islay	3,792	3,538	3,457	-8.8%
Tiree	Tiree	760	768	770	1.3%
Barra	Barra	1,264	1,244	1,078	-14.7%

Source: General Register Office for Scotland.

Table 2.2 Air and surface journey times on selected city-pair markets (in hours), July 2008

		Surface			
	Air	**Rail**	**Sea**	**Coach**	**Total**
Stornoway–Glasgow*	1.0	3.8	2.8	1.5	8.1
Orkney–Inverness**	0.8		0.7	4.0	4.7
Shetland–Edinburgh***	1.5	2.5	12.0		14.5
Wick–Aberdeen****	0.6	5.4			5.4

*Surface transport alternative for Glasgow–Stornoway: Glasgow–Inverness (rail), Inverness–Ullapool (coach), Ullapool–Stornoway (ferry). **Surface transport alternatives for Orkney–Inverness: Kirkwall–Burwick (bus), Burwick–John O'Groats (ferry), John O'Groats–Inverness (coach). ***Surface transport alternatives for Shetland–Edinburgh: Lerwick–Aberdeen (ferry), Aberdeen–Edinburgh (rail). ****Surface transport alternatives for Wick–Aberdeen: Wick–Inverness–Aberdeen (rail)

Sources: OAG, Traveline.

History

The history of air transport provision in the Highlands began with pioneering flights during the 1930s operated by Scottish and Highland Airways, both carriers subsequently taken over by British European Airways (BEA) in the late 1940s. The formation of British Airways in 1974, when BEA merged with the British Overseas Airways Corporation (BOAC), had little immediate effect on the pattern of air services. The first initiatives to rationalise air services were taken in 1981 with the formation of the British Airways Highland Division as a profit centre within the airline. These developments coincided with deregulation in the UK domestic market and the emergence during the 1980s and 1990s of Loganair as a rival to British Airways in the Highlands and Islands. Due to low traffic volumes on most city-pair markets, head-to-head competition between carriers was both limited in scope and occasional in frequency. Competition lasted until 1994 when British Airways signed a franchise agreement with Loganair and the Isle of Man-based British Regional Airlines (subsequently renamed BRAL and then BA CitiExpress) to operate services on its behalf in the Highlands and Islands. This initiative was motivated firstly by mounting financial losses the airline was incurring on directly operating air services in the region and, secondly, by a belief that franchising would enable carriers better suited to operating in such an environment to manage these services at lower cost while at the same time maintaining, and in some cases extending, the British Airways brand. BA CitiExpress, lured by the opportunities of developing in other more lucrative UK regional markets, eventually handed over all its internal Scottish services inherited from British Airways to Loganair in 2004.

Current Air Service Provision

At present there are both subsidised (PSO) and non-subsidised scheduled services within and to the Highlands and Islands. Loganair currently operates most of the non-subsidised services under a Flybe franchise agreement. The earlier agreement with British Airways Express was terminated in October 2008. Humberside-based airline Eastern Airways operates flights between Aberdeen and Wick and Aberdeen and Stornoway, while Highland Airways offers a daily connection between Inverness and Stornoway.

Many of these services also face competitive pressures from subsidised passenger ferry services, such as the Northlink services between Aberdeen and the Orkney and Shetland Islands and the state-owned Caledonian MacBrayne services that connect the mainland with several Hebridean islands off the west coast.

The map in Figure 2.1 shows the present structure of the Highlands and Islands network. There have been no significant alterations to the shape and scope of the air service network. The only significant recent development was the launch of new PSO subsidised air services linking the newly constructed airfields of Coll and Colonsay with Oban.

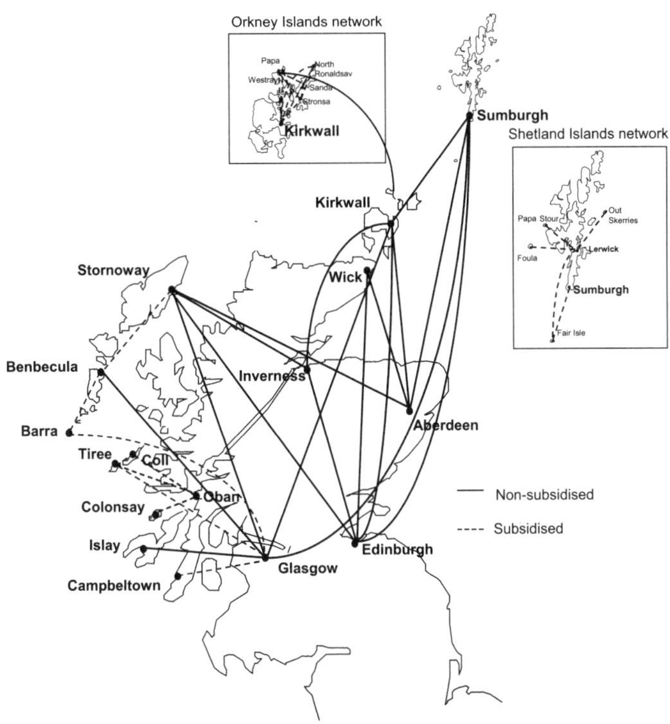

Figure 2.1 The Highlands and Islands air service network (July 2008)

Table 2.3 lists 2008 traffic data and weekly service frequency offered on each route in July 2008. It should be noted that on some routes operated by Loganair (Inverness–Kirkwall–Sumburgh, Sumburgh–Edinburgh) there has been a significant reduction in capacity offered as a result of high fuel prices.

Table 2.3 Total scheduled passenger traffic (2008) and weekly service frequency (July 2008) by route

	Total traffic	Weekly frequency (one-way)
Aberdeen–Sumburgh	68,357	29
Glasgow–Stornoway	58,144	22
Aberdeen–Kirkwall	44,177	25
Edinburgh–Kirkwall	35,656	19
Inverness–Stornoway	35,636	24
Edinburgh–Sumburgh	30,801	6
Glasgow–Islay	29,338	10
Edinburgh–Stornoway	28,320	15
Glasgow–Benbecula	25,201	11
Inverness–Kirkwall	25,085	6
Glasgow–Sumburgh	17,227	6
Aberdeen–Wick	16,539	15
Glasgow–Kirkwall	15,055	5
Glasgow–Campbeltown	9,072	10
Glasgow–Barra	9,066	4
Kirkwall–Sumburgh	8,709	6
Glasgow–Tiree	7,894	2
Inverness–Edinburgh	7,510	10
Stornoway–Benbecula	7,330	10
Edinburgh–Wick	7,255	5
Aberdeen–Stornoway	7,113	5
Barra–Benbecula	1,656	4
Inverness–Sumburgh	158	6
Inverness–Benbecula	18	5

Source: UK Civil Aviation Authority (2008); OAG (2008).

Regulation and Subsidy Policy

The provision of air transport in the Highlands and Islands is subject to the same licensing, safety and economic regulations that apply to other regions of the United Kingdom and the European Union. However, given the unremunerative nature of air service provision to remote regions and driven by the political and socio-economic need to integrate these communities, the UK government, under Article 4 of Regulation 2408/92 (Commission of the European Communities, 1992), imposed PSOs on a number of routes within the region. Public service obligations require a contracted air carrier to meet certain minimum service requirements in return for financial compensation over a fixed period of time. PSOs are currently imposed on 18 routes: Benbecula–Barra, Glasgow–Barra, Glasgow–Campbeltown, Glasgow–Tiree, Kirkwall–Eday, Kirkwall–Sanday, Kirkwall–Stronsay, Kirkwall–Westray, Kirkwall–Papa Westray, Kirkwall–North Ronaldsay, Lerwick–Foula, Lerwick–Papa Stour, Lerwick–Out Skerries, Lerwick–Fair Isle, Oban–Coll, Oban–Colonsay, Oban–Tiree and Stornoway–Benbecula.

Public service obligations are administered by municipal authorities on the three intra-island networks (Orkney, Shetland, Western Isles) and on the three Oban PSOs. The Scottish government administers PSOs on the routes from Glasgow. Traditionally Loganair was the dominant operator of PSO contracts in Scotland but in recent years new carriers have emerged, such as Highland Airways and Direct Flight. Highland Airways operates the Oban and Western Isles PSOs, while the Shetland intra-island service contract was awarded to Direct Flight.

The Scottish government also administers an Air Discount Scheme, which was set up in May 2006 in order to reduce the cost of air travel for residents living in Orkney, Shetland, the Western Isles, Islay, Jura, North West Sutherland and Caithness. These residents qualify for a 40 per cent discount on air fares to and from Aberdeen, Edinburgh, Glasgow and Inverness. The scheme, which was approved by the European Commission and will function until March 2011, does not apply to existing PSO services.

Regulatory Change and Services to Small Communities

United States

The United States was the first country in the developed world to liberalise its domestic scheduled airline market. The 1978 Airline Deregulation Act removed the structure of economic controls that had regulated the provision and pricing of scheduled air services since 1945.

By the mid to late 1980s there was growing concern amongst policy-makers in the United States that airline strategic decisions in response to new opportunities offered by deregulation were adversely impacting the level of air service provision to small communities. Evidence from studies undertaken during this period identified shifts in airline operating priorities, with some carriers exiting or reducing services within and to small community markets in favour of more financially lucrative higher-density routes. As a consequence of these developments, many non-hub airports lost direct services. Research by Jones and Cocke (1984), Molloy (1985) and Khil (1988) showed that the number of direct point-to-point services at non-hub airports declined markedly following deregulation in 1978. These arguments are, however, countered by Morrison and Winston (1986) and Butler and Huston (1990), who claim that air service provision to non-hub airports has actually improved since deregulation. This was manifested in increased flight frequencies on routes to small hubs and a greater range of destinations available to passengers through hub and spoke scheduling.

Another effect of deregulation was the trend amongst carriers to downsize aircraft. In liberalised markets, where airlines are competing for market share on the basis of improving service frequency, load factors are optimised through reducing aircraft size. Research by Huston and Butler (1988) showed that the percentage of turboprops deployed on the smallest city-pair markets increased from 52 per cent in 1976 to 91 per cent in 1988. This points to a shift in use from jets to turboprop aircraft, which many passengers perceive as offering a lower quality of in-flight service due to increased levels of noise and vibration compared to jet aircraft.

Europe

In the European Union (EU), the complete liberalisation of air service markets was not achieved until April 1997. Today a licensed air carrier can legally operate between any two airports open to scheduled services in the European Union outside those routes protected by public service obligations.

Studies by the European Commission (Commission of the European Communities, 1997) and the United Kingdom Civil Aviation Authority (UK Civil Aviation Authority, 1995) have shown that air travellers within the European Union have largely benefited from deregulation. These measures encouraged carriers to compete with established airlines in previously oligopolistic and monopolistic city-pair markets. Deregulation has also spawned the emergence of so-called low-cost 'no frills' airlines offering a greater choice of discount fares than was previously available. However, in evaluating the effects on a route-by-route basis, results are disappointing, with meaningful and effective head-to-head competition restricted to a small number of highly trafficked city-pairs (Commission of the European Communities, 1997). In light of these conclusions, one may ask to what extent has EU liberalisation benefited Europe's peripheral regions where traffic levels are significantly lower and air transport provision much more essential to social and economic development.

Research by Reynolds-Feighan (1995), who sought to identify the effects of liberalisation on traffic levels at small community Irish airports, found that liberalisation created an unstable operating environment, characterised by cyclical and permanent reductions in air service provision. The author cites the example of intense competition between Ryanair and Aer Lingus during the early 1990s, which led to the eventual withdrawal of direct services from Irish regional airports to both London and Dublin.[1] In citing this example, the author argues for a EU-wide essential air services programme of financial assistance along the same lines as that in place within the United States to protect air services to small and remote communities.

In Norway, also a country heavily dependent on air transport, domestic air service deregulation was initiated in 1975 with the partial relaxation of market entry restrictions and the certification of new scheduled air carriers. According to Ludvigsen (1993), small communities benefited from these measures. This was manifested in the subsequent launch of new intra- and inter-regional domestic services. The Norwegian government was restrained by socio-economic policy considerations from undertaking a process of complete liberalisation of the domestic market. This was due to concerns that deregulation followed by successful market entry would undermine the geographic integrity of the SAS and Braathens SAFE air carrier networks, which allowed for internal cross-subsidy between trunk routes and unprofitable services to remote communities. This ability to cross-subsidise therefore obviated the need for government to directly support services to airports located in remote regions.

Franchising Air Services

Air service franchising has been adopted by a growing number of airlines both in Europe and in North America over the last fifteen years. Motivated by the need to reduce and minimise costs in a highly competitive operating environment, several large carriers have out-sourced segments of their route networks to local franchise operators.

Franchisee carriers, according to Denton and Dennis (2000), as well as receiving the benefits of being aligned to a major airline brand, are also able to secure access to the global network, computer reservation system and the frequent flier programme of the franchiser and its other alliance partners. Franchising in Europe was pioneered by British Airways, where, under the British Airways Express brand, several UK-based carriers (BA CitiExpress, Loganair) operated regional services on behalf of the airline. Franchising also provides an opportunity for a large airline to access a market at minimal cost through establishing an agreement whereby a regional airline directly operates services on its behalf. Interestingly British Airways, as part of its new strategy, has since terminated most of its regional franchise agreements. However, in spite of this, franchising still remains popular with major carriers such as Air France and Lufthansa.

---

1 Galway, Carrickfinn, Sligo, Kerry (since reinstated as PSO routes) and Waterford to London and the reduction in Aer Lingus feeder flights from these points to Dublin.

It is apparent that there are mutual benefits to both the franchiser and the franchisee. However, to what extent has franchising affected the availability of convenient services and the quality of air service provision in general? One possible effect of out-sourcing existing services to a franchise operator is that the franchise operator is unable to match levels of in-flight service and punctuality previously set by the franchiser. The extent to which one is able to identify if such an effect took place is rather limited since no research has been undertaken to validate this view. However, the British Airways franchisee BRAL was heavily criticised by the north of Scotland press over poor punctuality not long after inheriting some Scottish internal routes from British Airways (Press and Journal, 1999).

On the other hand, it is possible that service levels in terms of increased frequency could improve with the deployment of a more appropriate mix of aircraft. A report published by the UK Civil Aviation Authority on trends emerging from the single aviation market after five years (UK Civil Aviation Authority, 1998) showed that markets previously not served could be served if operated by a regional franchise carrier. Robertson (1997) in an analysis of the effects of franchising on UK domestic routes, found that significant increases in traffic were recorded on those routes that became franchised and that a number of city-pairs currently without a direct service could potentially support a service if it were operated as a franchise. Denton and Dennis (2000), in their analysis of the impact of franchising on airlines and consumers, conclude that franchising can expand the scope and provision of air services in terms of routes to new destinations and increase frequency. However, the authors also note that the impact may also be adverse from a consumer's perspective since there is evidence, validated by research undertaken by the CAA (UK Civil Aviation Authority, 1998), that franchising has reduced competition within Europe's domestic markets while lower costs incurred by the franchisee have yet to feed through into lower fares for the consumer.

Measuring Changes in Levels of Air Service

One approach to measuring the impact of franchising is to develop indicators of service level and compare and contrast these over a specific time period. As franchising occurred after 1994 it is possible to contrast service levels both before and after this important milestone. Analysis is based on contrasting the level and quality of service provision at specific intervals in the period 1983 to 2006 using the following indicators:

- *Frequency*. Measures of total network frequency and average frequency per route were obtained. Statistics were complied from the *ABC/OAG World Airways Guide* for the month of June in the following years: 1983, 1988, 1993, 1998, 2001 and 2006.

- *Maximum lapsed destination time.* Maximum lapsed destination time (MLTD) measures the total lapsed time between the earliest daily arrival time and the latest daily departure time at a given location on a specific route. This variable is able to measure the relative convenience of departure and arrival times from the perspective of the day-trip passenger. Statistics were compiled from the same source as for weekly service frequency.
- *Seating capacity per flight.* Past research has shown that most passengers prefer travelling on jet aircraft as opposed to turboprops, and larger aircraft as opposed to small (Kanafani and Ghobrial, 1985; Ghobrial, 1989). Average seating capacity per flight measures the size of aircraft deployed on a given route and is used to identify if a specific market has experienced an increase or decrease in average aircraft size. Statistics were compiled from the same source as for weekly service frequency.
- *In-flight time.* In-flight time measures the amount of time spent by a passenger in flight on a specific route. This is determined by published scheduled departure and arrival times. Where there are different aircraft types deployed on a route, an average is obtained, weighted by the service frequency associated with each aircraft type.

Analysis is also focussed on assessing changes in patterns of air services within the context of whether franchise carriers were better able to manage capacity in relation to traffic.

The sample of routes selected for analysis, including sector distances, are listed in Table 2.4. Only those routes that were directly affected by franchising and were operated commercially (non-PSO) throughout the entire period were considered for analysis.

Table 2.4 Sector distances (in miles): sample routes

Aberdeen–Kirkwall	124
Aberdeen–Sumburgh	198
Edinburgh–Kirkwall	209
Edinburgh–Wick	174
Glasgow–Benbecula	158
Glasgow–Islay	72
Glasgow–Kirkwall	221
Glasgow–Stornoway	178
Inverness–Kirkwall	107
Inverness–Stornoway	96
Kirkwall–Sumburgh	94

Source: ABC (1993).

Results

Weekly Service Frequency

One-way service frequency aggregated across all routes increased by a factor of 65 per cent between 1983 and 2006, from 110 to 180 weekly flights. According to Figure 2.2 average weekly frequency per route, over the same period, increased from 10 to 16 flights per week. Average weekly frequency weighted by route density increased from 13 to 21 services per week during the same period.

Between 1988 and 1993 there was a 19 per cent increase in weighted average frequency per route. This was as a result of Loganair's expansion of services between 1991 and 1993 and an increase in frequency supplied by British Airways on the Aberdeen–Sumburgh, Glasgow–Stornoway and Inverness–Stornoway routes. This was followed by a modest contraction in service frequency across the network between 1993 and 1998, which coincides with the immediate effects of franchising. However, a pronounced expansion occurred between 2001 and 2006 when there was a 45 per cent increase in average service frequency, primarily on some of the busiest routes in the network such as Inverness–Stornoway (136 per cent), Glasgow–Stornoway (100 per cent) and Glasgow–Benbecula (117 per cent). For example, in June 2006, Loganair was offering four weekday rotations between Glasgow and Stornoway in contrast to the double daily services offered prior to franchising. Similarly on Aberdeen–Sumburgh, British Airways operated three weekday rotations in June 1983 compared to Loganair's five daily rotations in June 2006.

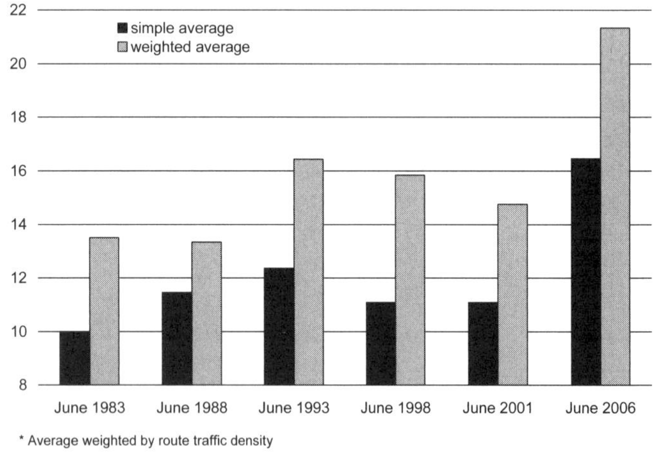

Figure 2.2 Simple and weighted* average weekly service frequency (one-way) per route, 1983–2006

Sources: ABC (1983, 1988, 1993); OAG (1998, 2001, 2006).

Evaluating individual route service frequency trends in Table 2.5 reveals two distinct phases of development since franchising was adopted. In the period 1993–2001, there was a 10 per cent contraction across the network in service frequency due to franchising. Only three of the selected routes recorded an increase in service frequency over the period (Edinburgh–Kirkwall, Aberdeen–Kirkwall and Inverness–Kirkwall). In contrast, significant reductions in frequency were recorded on Glasgow–Kirkwall, Glasgow–Stornoway and Kirkwall–Sumburgh, while frequency on Inverness–Kirkwall increased by a factor of 50 per cent.

The second phase of development occurred post-2001, when there was a significant increase in total service frequency from 122 weekly flights in June 2001 to 188 flights in June 2006. This period coincided with Loganair replacing BRAL on three British Airways franchise routes (Glasgow–Stornoway, Aberdeen–Sumburgh, Glasgow–Benbecula). Frequency increased on all but three routes, with the most impressive growth rates achieved on Glasgow–Benbecula, Inverness–Stornoway, Glasgow–Stornoway and Kirkwall–Sumburgh. On Glasgow–Stornoway, BRAL's twice-daily service operated by 66-seat BAe ATP aircraft was replaced by smaller Loganair Saab 340s operating a weekday frequency of four round trips.

However, one disadvantage in adopting frequency as a measure of the level and quality of service offered by an airline is that little indication is provided as to the convenience to passengers of departure and arrival times on a given route.

Table 2.5 Weekly service frequency (one-way) by route in June 1983, 1988, 1993, 1998, 2001 and 2006

	1983	1988	1993	1998	2001	2006	% change 1993–2006
Aberdeen–Kirkwall	11	16	17	20	18	24	41%
Aberdeen–Sumburgh	23	18	25	24	22	33	32%
Edinburgh–Kirkwall	5	11	6	5	12	10	67%
Edinburgh–Wick	5	11	6	6	6	5	-17%
Glasgow–Benbecula	6	7	6	6	6	13	117%
Glasgow–Islay	9	11	11	11	11	15	36%
Glasgow–Kirkwall	6	5	21	7	6	7	-67%
Glasgow–Stornoway	11	12	17	14	12	24	41%
Inverness–Kirkwall	15	16	8	10	12	12	50%
Inverness–Stornoway	8	8	11	12	11	26	136%
Kirkwall–Sumburgh	11	11	8	7	6	12	50%

Sources: ABC (1983, 1988, 1993); OAG (1998, 2001, 2006).

Maximum Lapsed Destination Time

Schedule convenience can be measured by calculating maximum lapsed destination time (MLDT) in both directions for each city-pair. MLTD measures the total lapsed time between the earliest daily arrival time and the latest daily departure time at a given location on a specific route.

Figure 2.3 shows that there has been a steady and consistent improvement in both simple and weighted average MLDT over the entire period. However, what is particularly noticeable is the increase recorded between 2001 and 2006. Over the entire period weighted average MLDT per route increased quite markedly from 5.58 hours in 1993 to 10.21 hours in 2006, while the simple average increased from 5.44 hours to 8.99 hours over the same period.

Observation of the statistics listed in Table 2.6 shows that there is significant variation in absolute levels of MLDT between and within city-pair markets, in the latter case through directional differences in absolute levels and in growth rates between each period. To a large extent, differences within and between city-pairs are as a result of scheduling and aircraft overnight positioning decisions taken by airlines. From 1983 to 1993, there was little change in MLDT, with a few exceptions. For example, on services from Aberdeen to Kirkwall, MLDT increased by 67 per cent.

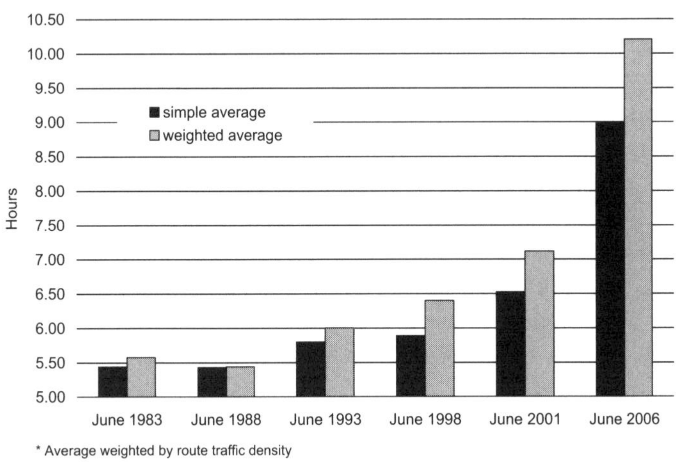

Figure 2.3 Simple and weighted* average MLDT (in hours) per route, 1983–2006

Sources: ABC (1983, 1988, 1993); OAG (1998, 2001, 2006).

Table 2.6 Maximum Lapsed Destination Time in hours (MLDT) by route in June 1983, 1988, 1993, 1998, 2001 and 2006

	June 1983	June 1988	June 1993	June 1998	June 2001	June 2006	% change 1993–2006
Sumburgh–Aberdeen	6.83	6.50	7.17	9.50	9.58	9.50	33%
Aberdeen–Sumburgh	6.17	5.25	5.58	6.33	6.25	12.50	124%
Kirkwall–Aberdeen	6.83	6.92	7.17	6.75	6.00	7.83	9%
Aberdeen–Kirkwall	3.58	4.25	6.00	9.25	9.25	10.42	74%
Glasgow–Islay	7.58	7.25	7.33	8.58	8.33	9.08	24%
Islay–Glasgow	5.92	5.50	5.58	6.58	6.33	4.92	-12%
Glasgow–Stornoway	3.42	3.83	7.42	3.17	6.92	10.50	42%
Stornoway–Glasgow	5.42	5.50	4.50	4.67	3.83	7.50	67%
Inverness–Kirkwall	3.67	4.25	4.33	5.83	5.25	10.08	133%
Kirkwall–Inverness	6.25	5.92	5.92	8.83	9.50	9.42	59%
Kirkwall–Sumburgh	1.33	1.08	1.25	4.42	4.42	4.83	287%
Sumburgh–Kirkwall	9.75	10.25	9.92	0.00	0.00	10.58	7%
Inverness–Stornoway	8.25	8.42	7.42	7.92	9.00	10.58	43%
Stornoway–Inverness	1.17	1.08	1.58	0.58	6.67	8.17	416%

Sources: ABC (1983, 1988, 1988b, 1993, 1993); OAG (1998, 1998, 2001,2006).

After 1993, following the franchising of air services, significant changes were apparent across most routes up to 2001. Large increases were recorded on Stornoway–Inverness (321 per cent). Only four out of the 14 directional routes recorded reductions in MLDT: Kirkwall–Aberdeen, Glasgow–Stornoway, Sumburgh–Kirkwall and Stornoway–Glasgow. The impact of franchising, therefore, on MLDT appears largely to have been positive. This was especially so after 2001 when Loganair replaced the three BRAL-operated routes. Growth was particularly impressive on Aberdeen–Sumburgh and Glasgow–Stornoway, both routes previously operated by BRAL in 2001. One important factor that contributed to the improvement in MLDT was the extension to airport opening hours post-2001, which enabled not only additional frequency but also much earlier and later departure and arrival times at these remote airports, thus allowing island residents the opportunity to spend additional time in Aberdeen, Edinburgh and Glasgow or to connect with a greater range of flights to other UK airports and beyond. For example, in June 2001 the earliest weekday arrival in Glasgow available to Stornoway residents was 11:40 and the latest departure back home on the same day was 15:30. In 2006, the improved Loganair schedule allowed for an early 09:30 arrival in Glasgow and a latest departure time for the return flight of

17:00. Significant improvements were also recorded on Aberdeen–Sumburgh and Inverness–Kirkwall.

Average Seating Capacity per Flight

In the Highlands and Islands, because of low traffic volumes all internal air services have been and still are operated by turboprop aircraft. Figure 2.4 shows that average seating capacity per flight increased from 36.1 in June 1983 to 52.1 in June 1993. Up to 1991, British Airways services were operated by 44-seat British Aerospace (BAe) 748 turboprops. These were then replaced by 66-seat BAe advanced turboprops (ATPs). This explains the sharp increase in both simple and weighted average aircraft size between 1988 and 1993. Rival carrier Loganair, prior to 1994, was operating a mix of de Havilland Twin Otters, BAe Jetstream 31s, Shorts 360s and Britten-Norman Islanders, all aircraft with fewer than 40 seats.

When British Airways services were franchised to BRAL and Loganair, 10 out of the 11 routes sampled recorded a reduction in average seating capacity per flight of 4–47 per cent between 1993 and 2001. For example, according to Table 2.7, average aircraft seating capacity per flight fell by 47 per cent between 1993 and 2001 on services between Aberdeen and Kirkwall. This was because BRAL replaced ATP turboprops, previously operated by British Airways, with smaller 34-seat Saab 340s. Similar scale reductions were recorded on other routes where ATP turboprops were replaced by older non-pressurised 34-seat Shorts 360 aircraft. These results suggest that service levels from the perspective of aircraft size have declined in so far as passengers travelling within and to and from the

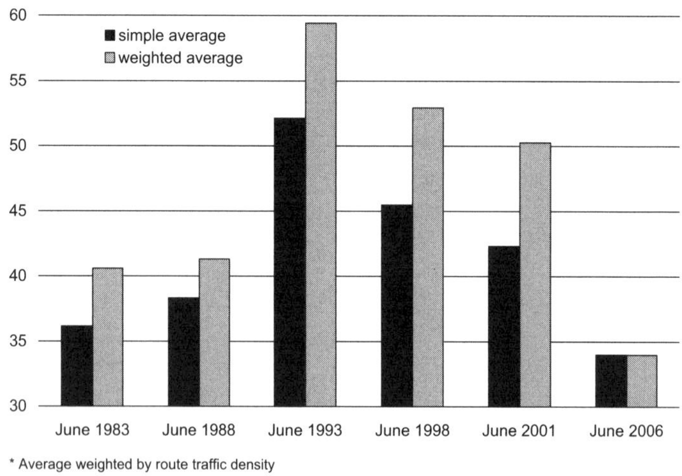

Figure 2.4 Simple and weighted* average aircraft seat capacity per flight per route, 1983–2006

Sources: ABC (1983, 1988, 1993); OAG (1998, 2001, 2006).

Highlands and Islands have had to endure the discomfort of not only smaller but, in some cases, non-pressurised aircraft as a direct consequence of franchising. However, it should be pointed out that between 1998 and 2001, Loganair was in the process of replacing its ageing non-pressurised Shorts 360 fleet with faster and pressurised Saab 340 turboprops. By June 2006, with BRAL having withdrawn its ATP operations and exited the market, all of the 11 routes were operated by 34-seat Loganair Saab 340s. The advantage here from the operators' perspective is the economies achieved through standardising and limiting the range of aircraft equipment to one type. For passengers, the advantage is that the operator is able to serve the market at a higher volume of service frequency at commercially sustainable load factor levels.

In-flight Time

Passenger perception of the quality of service offered in a given market is, to a large extent, influenced by the amount of time spent in flight between any two airports. Figure 2.5 illustrates the changes in simple and weighted average in-flight time per route between 1983 and 2006. In June 1983, average in-flight time was 0.78 hours per route. By June 1988, this had increased marginally to 0.82 hours. As far as weighted average in-flight time is concerned, there was no change between June 1983 and June 1988. Between 1988 and 1993, both simple and weighted average in-flight time declined to 0.72 hours and 1.02 hours respectively. Up to 1993, the deployment of new and faster aircraft by both Loganair and British Airways was largely responsible for the decline in both average and weighted average in-flight time.

Table 2.7 Average seating capacity per flight by route in June 1983, 1988, 1993, 1998, 2001 and 2006

	1983	**1988**	**1993**	**1998**	**2001**	**2006**	**% change 1993–2006**
Aberdeen–Kirkwall	44	44	64	35	34	34	-47%
Aberdeen–Sumburgh	44	44	64	61	62	34	-47%
Edinburgh–Kirkwall	20	35	36	34	34	34	-6%
Edinburgh–Wick	20	35	36	34	34	34	-6%
Glasgow–Benbecula	44	44	64	66	66	34	-47%
Glasgow–Islay	20	20	36	34	34	34	-6%
Glasgow–Kirkwall	44	44	37	39	34	34	-8%
Glasgow–Stornoway	44	66	64	63	66	34	-47%
Inverness–Kirkwall	30	24	45	34	34	34	-24%
Inverness–Stornoway	44	44	64	63	34	34	-47%
Kirkwall–Sumburgh	44	44	64	39	34	34	-47%

Sources: ABC (1983, 1988, 1993); OAG (1998, 2001, 2006).

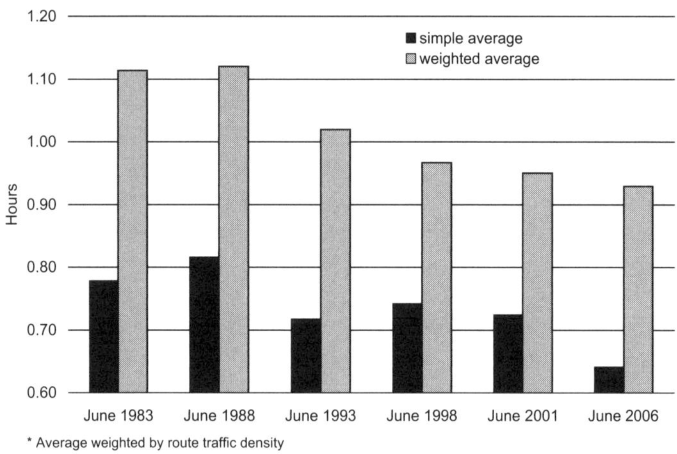

Figure 2.5 Simple and weighted* average in-flight time (in hours) per route, 1983–2006

Sources: ABC (1983, 1988, 1993); OAG (1998, 2001, 2006).

Franchising appears to have had the effect of reducing further, with the exception of a very slight increase in the simple average between 1993 and 1998, average in-flight time across the network. These trends occurred in spite of the fact that slower Shorts 360 aircraft were being deployed in place of faster ATPs on a number of routes (Inverness–Kirkwall, Edinburgh–Kirkwall, Kirkwall–Sumburgh and Glasgow–Kirkwall). The decline in average in-flight time was mainly due to an increase in the number of non-stop direct flights across the network. Figure 2.6 shows that the number of direct non-stop services as a percentage of total direct services declined from 74 per cent in June 1983 to 68 per cent in June 1993. However, as smaller aircraft were introduced to the network after franchising a greater number of direct services were introduced, increasing the proportion to 85 per cent in June 1998.

Analysis of city-pair in-flight time data listed in Table 2.8 illustrates the combined effect of replacing direct with one-stop flights, and vice versa, and reducing aircraft size on in-flight time. For example, on services between Aberdeen and Kirkwall, the replacement of the 66-seat BAe ATP with the 34-seat Saab 340 operated by BRAL did not affect in-flight time since the Saab 340, despite its lower seating capacity, is able to match the cruising speed achieved by the ATP. Furthermore, in 1993 five of the 17 weekly flights between Aberdeen and Kirkwall involved an intermediate stop in Inverness, whereas in 1998 and 2001 all flights between both airports were non-stop.

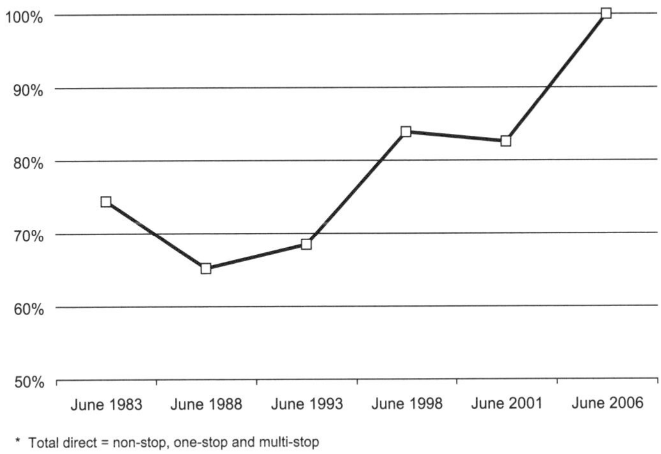

Figure 2.6 Direct non-stop services as a percentage of total direct services,* 1983, 1988, 1993, 1998, 2001 and 2006

Sources: ABC (1983, 1988, 1993); OAG (1998, 2001, 2006).

Table 2.8 Average in-flight time (in hours) by route in June 1983, 1988, 1993, 1998, 2001 and 2006

	1983	1988	1993	1998	2001	2006	% change 1993–2006
Aberdeen–Kirkwall	0.75	1.10	0.99	0.75	0.83	0.83	-16%
Aberdeen–Sumburgh	1.49	1.34	1.13	1.00	1.00	1.00	-12%
Edinburgh–Kirkwall	1.83	2.18	1.46	2.01	1.77	1.33	-9%
Edinburgh–Wick	1.33	1.58	1.01	1.17	1.08	1.00	-1%
Glasgow–Benbecula	1.00	0.92	1.00	1.00	1.00	1.00	0%
Glasgow–Islay	0.67	0.58	0.58	0.58	0.67	0.75	29%
Glasgow–Kirkwall	2.08	2.00	1.38	2.02	2.14	1.00	-4%
Glasgow–Stornoway	1.08	1.09	1.27	1.13	1.00	0.75	-21%
Inverness–Kirkwall	0.95	1.01	0.67	0.88	0.75	0.67	13%
Inverness–Stornoway	0.67	0.67	0.67	0.67	0.67	0.58	0%
Kirkwall–Sumburgh	0.58	0.58	0.58	0.67	0.67	0.83	0%

Sources: ABC (1983, 1988, 1993); OAG (1998, 2001, 2006).

In-flight time increased by the largest amount between 1993 and 2001 on Glasgow–Kirkwall and Edinburgh–Kirkwall. Prior to franchising, Loganair operated direct services on these routes in competition with British Airways using newly acquired British Aerospace Jetstream 31s. However, as a result of the restructuring that followed the franchise agreement, both routes were eventually handed over to BRAL in 1994, which replaced direct non-stop services with one-stop flights operated by non-pressurised Shorts 360s involving intermediate stops at Inverness and Aberdeen. The effect of this was to increase average in-flight time on Edinburgh and Glasgow services respectively. However, services between Aberdeen and both Sumburgh and Kirkwall experienced a reduction in flight time due to the replacement of one-stop services with non-stop flights. Despite the substitution of ATPs with the smaller Saab 340 on the Kirkwall–Aberdeen route, there was no adverse effect on in-flight time since the Saab 340 is able to match the cruising speed achieved by the ATP.

Franchising, which coincided with the deployment of smaller aircraft on the Highlands and Islands network, appears to have been responsible for the overall reduction in flight time. Before 1993, British Airways was operating 66-seat ATPs which could only be operated commercially on a multi-sector basis, hence the low proportion of direct non-stop services in June 1993. However, the franchise carriers, especially Loganair, had a more appropriate mix of equipment for the market and were thus able to offer more direct services and a reduction in in-flight time.

There was a further improvement in average in-flight time after 2001. A combination of the phasing out of the slower Shorts 360 and the substitution of multi-stop services with direct connections helped contribute to a further 11 per cent reduction in average flight time. In June 2006, all flights across the selected sample of routes were non-stop. This move towards direct non-stop services assisted by the use of smaller aircraft also may have had some unintended consequences. Inverness no longer functions as a transit stop for services between Glasgow and the Northern Isles. Because of the introduction of direct flights between Glasgow and Kirkwall and Glasgow and Stornoway there is less need to use Inverness, and hence the removal of twice-daily direct flights between Inverness and Glasgow in 2004. The demise of this route was also due mainly to the development of more direct connections between Inverness and several UK regional airports, which reduced the need for passengers to access these markets through connecting services at Glasgow. The loss of connecting traffic on this route prompted Loganair to withdraw the route.

Management of Supply and Demand

Without the benefit of government subsidies or cross-subsidisation from profitable routes, air carriers operating in low-density markets are driven by the commercial necessity of recovering costs and generating a return on their investment. This can pose a particular economic challenge in an environment of comparatively high infrastructure user charges and low traffic volumes. Therefore, an analysis of service provision would be incomplete without adequate consideration of the management and deployment of capacity in relation to developments in the volume of passenger traffic.

Table 2.9 provides a breakdown of seating capacity, passenger traffic levels and passenger load factors on routes within the Highlands and Islands. Statistics are separated into east coast and west coast services. Total passenger traffic on services in the Highlands and Islands increased from 17,821 in June 1983 to 31,053 in June 2001, an average annual growth rate of 4.1 per cent. This compares to an average annual growth rate of 2.7 per cent in seating capacity. Total load factor in June 1983 was 67 per cent, increasing to 73 per cent in June 1988.

Table 2.9 Passenger traffic, capacity and load factor by route sector in June 1983, 1988, 1993, 1998, 2001 and 2006

	1983	1988	1993	1998	2001	2006
Passenger traffic						
East coast*	11,858	12,406	16,526	19,465	21,909	20,886
West coast**	5,963	7,908	9,336	9,991	9,144	12,218
Total	17,821	20,314	25,862	29,456	31,053	33,104
Total seating capacity						
East coast*	16,192	15,712	26,256	24,960	26,896	30,736
West coast**	10,240	12,192	17,504	19,064	12,496	21,216
Total	26,432	27,904	43,760	44,024	39,392	51,952
Load factor						
East coast*	73%	79%	63%	78%	81%	68%
West coast**	58%	65%	53%	52%	73%	58%
Total	67%	73%	59%	67%	79%	64%

*Includes all services between Aberdeen and Orkney and Shetland in addition to all routes between Glasgow/Inverness and Orkney and Shetland.

**Includes all services between the Scottish mainland and the Western Isles.

Sources: ABC (1983, 1988, 1993); OAG (1998, 2001, 2006).

Between 1988 and 1993, the combined effect of additional Loganair services and the replacement of BAe 748 turboprops with larger ATPs on the British Airways network resulted in a reduction in load factor on both east and west coast route sectors. For example, load factor on services between the mainland (Glasgow, Inverness) and Western Isles (Stornoway, Benbecula, Islay) declined from 65 per cent in June 1988 to 53 per cent in June 1993. Load factor also fell on the east coast route network from 79 per cent to 63 per cent over the same period.[2]

What is also interesting is that levels of traffic and the deployment of capacity have not been affected by declining population levels of some of the island communities reported in Table 2.1 if one contrasts the situation in 2001 with 1983. However, it is important to note that traffic levels on some of the mainland–Western Isles routes have actually fallen between 1998 and 2001. For example, the total number of passengers carried on Glasgow–Islay in 2001 was 19,609, compared to 19,914 in 1998. This trend may be as a result of two factors: depopulation or changes in service levels and prices on competing ferry services.

Increased competition during the period 1991–94 between British Airways and Loganair and the deployment of large aircraft had the combined effect of creating an environment of excess capacity. For example, Loganair introduced direct services for the first time in June 1993 between Glasgow and Kirkwall, in competition with daily British Airways services which were routed through Aberdeen and Inverness. However, such a strategy could not be sustained in the face of mounting financial losses; hence, the commercial logic in seeking to co-operate rather than compete with rival Loganair.

Where competition, though limited, had existed before the effect of the franchise agreement was to make the British Airways brand dominant in the Highlands. The franchise agreement allowed the air carriers to establish a more financially sustainable network of services through adjustments in frequency, routings, capacity and aircraft type. The franchise carriers were able to achieve significantly higher load factors, 79 per cent in June 2001 compared to 59 per cent in June 1993. Indeed, load factors, which had been comparatively low on west coast routes between 1983 and 1998, had improved markedly by June 2001. The main reason for this is that Loganair had replaced BRAL on the Inverness–Stornoway route and were offering a reduced level of capacity (ATPs replaced by Saab 340s) but, interestingly, a more convenient schedule of services (see Table 2.6). What is apparent from these developments is that the rationalisation and consolidation that was inevitable in an environment of excess capacity was achieved without any significant deterioration in service levels. Indeed, many aspects of air service provision improved, such as the convenience of flight timings, reduced in-flight times and the introduction of new routes as a consequence of franchising. However the expansion in frequency and capacity post-2001 that also secured improved departure and arrival times for passengers resulted in a decline in average load

2 It should also be borne in mind that these statistics relate to the month of June and one would expect lower load factors between October and March.

factor from 79 per cent in 2001 to 64 per cent in 2006. Reductions were also recorded on both east and west coast route sectors. Loganair system-wide load factor also declined during this period from 62 per cent in June 2001 to 58 per cent in June 2006. This suggests that the market was over-supplied and that a downward correction over the medium term would be highly likely.

Future Implications

In October 2008, a franchise agreement with Flybe replaced the British Airways contract that had been in place with Loganair since 1994. One of the advantages of the British Airways franchise agreement was that it provided residents in very remote regions of Scotland with access to a global network of services offered by one airline. It was possible, for example, to make one booking between Islay and New York without the need for any change of airline. This arrangement though has continued in the form of a code-share agreement between Flybe and British Airways on Scottish routes operated by Loganair. The new arrangement will therefore ensure that residents in remote Highland communities remain connected to the British Airways global air services network. The risk, of course, is that this arrangement may be terminated at some stage in the future, which would have the effect of drastically reducing the range of international destinations available with seamless connections to airports in the Highlands and Islands, as Flybe is predominantly a UK domestic carrier offering a very limited range of international services.

One of the most significant and pressing business challenges likely to face Loganair will be the availability of a suitable aircraft type to replace the existing fleet of 16 34-seat Saab 340 turboprops, the majority of which are leased and were built in the late 1980s/early 1990s – operation of older aircraft increases maintenance costs. This and other comparably sized aircraft types such as the BAe Jetstream 41 (29 seats), Dornier 328 (31 seats) and Bombardier Dash-8-100 series (37 seats) are all no longer in production (see Chapter 16). The only aircraft type similar to the Saab 340 currently still in production is the 50-seat ATR-42. In order to maintain the commercial viability of existing routes, deployment of replacement aircraft with higher seating capacity would most certainly require a reduction in frequency, which could reverse many of the gains in service convenience that have been achieved across most routes in the Highlands and Islands network. One possible solution to dealing with the risk of service reduction would be to extend the PSO regime to other routes in the network. This would apply to those vulnerable routes that are at risk of a reduction in service frequency due to larger aircraft deployment.

Conclusion

Air services were franchised by British Airways in order to reduce costs and to eliminate the mounting losses incurred in operating services directly. Franchise carriers, through using a more appropriate size and mix of aircraft, were better able to achieve higher load factors and more commercially successful operations. In spite of the fact that passengers had to endure the relative discomfort and inconvenience of smaller aircraft as a result of downsizing, franchising has brought improved departure and arrival times, shorter journeys and fewer multi-stop services. Franchising has also resulted in new routes, such as Edinburgh–Stornoway, and a new direct non-stop service between Edinburgh and Sumburgh. From the operators' perspective the over-provision in air services that was clearly apparent in 1993 could not be sustained and the reduction in capacity was necessary in an environment of low traffic volumes and high costs.[3] These were the key factors that prompted British Airways to franchise its Highlands and Islands services to BRAL and Loganair. Both airlines were better able to serve the market through a combination of lower costs and more suitable aircraft types.

Although this analysis is rather limited in its geographic scope, it has sought to examine how franchising has impacted on the provision of air services in one of Europe's peripheral regions. It is also very important to point out that enhanced service levels have been achieved without the imposition of public service obligations with their fixed minimum levels of service and financial compensation.

References

ABC (1983), *World Airways Guide*. June (Dunstable: Reed Travel Group).

ABC (1988), *World Airways Guide*. June (Dunstable: Reed Travel Group).

ABC (1993), *World Airways Guide*. June (Dunstable: Reed Travel Group).

Butler, R. and Huston, J. (1990), 'Airline service to non-hub airports: ten years after deregulation', *Logistics and Transportation Review* 26:1, 3–17.

Commission of the European Communities (1992), *Council Regulation No 2408/92 on Access for Community Air Carriers to Intra-community Air Routes* (Brussels: Commission of the European Communities).

Commission of the European Communities (1997), *Single Market Review Impact on Services: Air Transport* (London: Kogan Page).

Denton, N. and Dennis, N. (2000), 'Airline franchising in Europe: benefits and disbenefits to airlines and consumers', *Journal of Air Transport Management*, 6, 179–90.

General Register Office for Scotland (2008), <http://www.gro-scotland.gov.uk/>.

3 British Airways, for example, was reported to have been losing approximately US$5 million per annum on services between the mainland and island communities.

Ghobrial, A. (1989), 'Competition between US and foreign carriers: an econometric approach', *International Journal of Transport Economics*, 16:1, 19–33.

Highlands and Islands Enterprise (2002), *Economic Information*. <http://www.hie.co.uk/>.

Huston, J. and Butler, R. (1988), 'The effects of fortress hubs and airline fares and service: the early returns', *Logistics and Transportation Review*, 24:3, 203–13.

Jones, R. and Cocke, S. (1984), *Community Air Services under Deregulation*. Presented at Southwest Economics Association Meeting.

Kanafani, A. and Ghobrial, A. (1985), 'Airline hubbing: some implications for airport economics', *Transportation Research*, 19:1, 15–27.

Khil, M. (1988), 'The impacts of deregulation of passenger transportation in small towns', *Transportation Quarterly*, April, 243–68.

Kyrou, D. (1998), *Memorandum to the House of Commons Environment, Transport and Regions Select Committee Report on Regional Air Services*. HC-589 (II).

Ludvigsen J. (1993), 'Liberalisation of market entry for Norwegian carriers', *Transportation Journal*, summer, 41–54.

Molloy J. (1985), *The US Commuter Airline Industry: Policy Alternatives* (Lexington: Heath and Company).

Morrison, S. and Winston, P. (1986), *The Economic Effects of Airline Deregulation* (Washington DC: Brookings Institute).

OAG (1998), *World Airways Guide*. June edition (Dunstable: OAG).

OAG (2001), *World Airways Guide*. June edition (Dunstable: OAG).

OAG (2006), *World Airways Guide*. June edition (Dunstable: OAG).

OAG (2008), *World Airways Guide*. July edition (Dunstable: OAG).

P&O Scottish Ferries (2002), *Sailing Schedules*. <http://www.poscottishferries.co.uk>.

Press and Journal (1999), 'Islands MSPs meet the airline bosses', 24 June, p. 3.

Reynolds-Feighan, A. (1995), 'European and American approaches to air transport liberalisation: some implications for small communities', *Transportation Research Part A: Policy and Practice*, 29:6, 467–83.

Robertson, N. (1997), *An Analysis of the Effects of Franchising on the UK Domestic and UK–EU Markets*. Unpublished MSc thesis (Cranfield: Cranfield University).

UK Civil Aviation Authority (1974), *Air Transport in the Scottish Highlands and Islands* (London: Civil Aviation Authority).

UK Civil Aviation Authority (1995), *The Single European Aviation Market: Progress So Far* (London: Civil Aviation Authority).

UK Civil Aviation Authority (1998), *The Single European Aviation Market: The First Five Years*, CAP 685 (London: Civil Aviation Authority).

UK Civil Aviation Authority (2001), *UK Airports: Annual Statements of Movements, Passengers and Cargo 2001*, CAP 662 (London: Civil Aviation Authority).

Western Isles (2002), *Western Isles Council European and Development Services*. <http://www.w-isles.gov.uk/factfile/population>.

Williams, G. and Pagliari, R. (2004), 'A comparative analysis of the application and use of public service obligations in air transport within the EU', *Journal of Transport Policy*, 11:1, 55–66.

Chapter 3
Developing Air Services in the Azores

Luís Silveira

The Azores Region

The Azores archipelago is one of Europe's remotest regions, located in the North Atlantic about 1,800 km from the Portuguese mainland and 4,500 km from North America at longitude 24°46′/31°16′ and latitude 39°43′/36°55′. Year-round direct flights link the islands with Lisbon (2 hrs 15 mins), London (3 hrs 30 mins), Frankfurt (4 hrs 30 mins), Stockholm (5 hrs 30 mins), Boston (5 hrs 30 mins) and Toronto (6 hrs). The nine islands that comprise the archipelago have a total land area of 2,333 km^2, with Corvo the smallest at 17 km^2 and 445 inhabitants and São Miguel the largest at 747 km^2 and 130,800 inhabitants. Table 3.1 lists the population size and land area of each of the islands. The distances between the islands vary from just 6 km to 600 km.

The Azores, with its own parliament and regional government, has some characteristics common to many other remote regions and especially archipelagos, namely sparsity, small population and a weak or fragile economy. As a peripheral region in a small country, only in the last 30 years has the Azores received significant investment in essential structures. Tables 3.2 and 3.3 provide some recent economic data for Portugal's mainland and its two archipelagos.

Table 3.1 Azores Islands – populations and land areas

Island	Land area (km^2)	Population
Corvo	17	445
Faial	173	15,072
Flores	142	3,967
Graciosa	61	4,748
Pico	448	14,666
Santa Maria	97	5,496
São Miguel	747	130,839
São Jorge	246	9,539
Terceira	402	55,252
Total	2,333	240,024

Table 3.2 Employment by sector (%)

	Azores	Madeira	Portugal
Primary	12.4	9.9	11.5
Secondary	25.5	27.0	30.8
Tertiary	62.1	63.1	57.7

Source: INE 2004

Table 3.3 GDP per capita by Portuguese region

Region	Index number
North	79
Centre	86
Lisbon	141
Alentejo	94
Algarve	103
Madeira	121
Azores	88

Source: INE 2004

The Azores and the Birth of Commercial Aviation

The geographical location of the Azores provided an ideal refuelling point for the pioneers of flight across the North Atlantic in the early part of the twentieth century. In the 1930s the islands served as mid-Atlantic stations for the first international airlines to launch commercial flights between Europe and North and South America with seaplanes, such as Lufthansa, Air France and Pan American.

The first runways were built during World War II by the allied forces on Terceira and Santa Maria. Santa Maria Airport was used for many years for technical and commercial stops by European and American airlines operating transoceanic commercial routes. Some of these international flights also carried domestic traffic between the Azores and the Portuguese mainland. With the increased technical capability of jet aircraft and the beginning of domestic flights by the Portuguese national carrier this situation came to an end. Azorean airports thus played an important role in supporting the development of transoceanic commercial aviation, not only in providing ground stations for technical and commercial stops but also in offering suitable alternative airports for jet aircraft operating transoceanic flights under Extended-range Twin-engine Operational Performance Standards (ETOPS) rules.

SATA, the Azorean airline, is one of the oldest in the world, beginning operations in June 1947. The airline's purpose was to provide connections between São Miguel, the largest island, and transatlantic international flights operating via Santa Maria. Over time, more of the islands were provided with the necessary airport infrastructure, with the number of inter-island flights increasing accordingly.

The Airport System in the Azores

Nowadays, all nine islands in the Azores have an airport. Table 3.4 provides details of the runways at each of the airports. Lajes Airport in Terceira was built during World War II and continues in use as a military facility for the Portuguese and US Air Forces. Commercial flights there are subject to military coordination and priorities. The other main airports are state owned and managed by ANA. Four airports were built by the regional government only at the beginning of the 1980s. Table 3.5 shows the current situation concerning ownership and management of airports in the Azores.

The airports on Santa Maria, São Miguel, Horta and Terceira are classified as international airports, but only Terceira and São Miguel currently receive international flights, as shown in Table 3.6. The composition of traffic at the different airports accordingly varies significantly, as is apparent from Table 3.7.

Table 3.4 Azorean airport runway details

Island (airport code)	Length (m)	Width (m)
Corvo (CVU)	800	20
Faial (HOR)	1,595	45
Flores (FLW)	1,400	30
Graciosa (GRW)	1,325	30
Pico (PIX)	1,745	45
Santa Maria (SMA)	3,048	60
São Miguel (PDL)	2,353	45
São Jorge (SJZ)	1,310	30
Terceira (TER)	3,314	60

Table 3.5 Azorean airport ownership and management

Airport	Facility	Ownership	Management
Corvo (CVU)	All	Regional authority	SATA subsidiary
Faial (HOR)	All	State	ANA
Flores (FLW)	Runway	State	ANA
	Terminal	Regional authority	SATA subsidiary
Graciosa (GRW)	All	Regional authority	SATA subsidiary
Pico (PIX)	All	Regional authority	SATA subsidiary
Santa Maria (SMA)	All	State	ANA
São Miguel (PDL)	All	State	ANA
São Jorge (SJZ)	All	Regional authority	SATA subsidiary
Terceira (TER)	Runway	Military	Portuguese and US Air Forces
	Terminal	Regional authority	Regional authority

Table 3.6 Types of passenger traffic at Azorean airports

Airport	Types of passenger traffic
Corvo (CVU)	Inter-island only
Faial (HOR)	Inter-island and Portuguese mainland
Flores (FLW)	Inter-island only
Graciosa (GRW)	Inter-island only
Pico (PIX)	Inter-island and Portuguese mainland
Santa Maria (SMA)	Inter-island and Portuguese mainland
São Miguel (PDL)	All
São Jorge (SJZ)	Inter-island only
Terceira (TER)	All

Table 3.7 Share of passenger traffic by airport (%)

Airport	% share of 1.75 m passenger traffic
Corvo (CVU)	0.2
Faial (HOR)	10.7
Flores (FLW)	2.1
Graciosa (GRW)	1.9
Pico (PIX)	2.7
Santa Maria (SMA)	5.2
São Miguel (PDL)	49.5
São Jorge (SJZ)	2.6
Terceira (TER)	25.2

Inter-island Market Structure

Until 2003, Azorean airports were exempt from the European Union's Third Package Market Access Rules (EEC Regulation no. 2408/92). As a consequence, SATA continued to operate the inter-island services under a monopoly concession until March 2003. Since then services have been subject to Public Service Obligation (PSO) legislation and are operated under a concession contract for three-year terms following an international public tender. The PSO contract was re-tendered in 2003, 2006 and 2009, and in all instances SATA was the only airline to submit a bid. The PSO conditions for the air services in the Azores are quite demanding.[1] In all there are 15 inter-island PSO routes, some of which are just seasonal. Figure 3.1 shows the routes and respective traffic flows in 2006.

Most of the inter-island traffic comprises resident passengers who in winter months account for up to 70 per cent of demand. The annual average however is 60 per cent residents and 40 per cent non-residents, resulting from the effects of tourism in the summer season. Some routes are very thin in terms of demand, in accordance with respective island size and population as per Figure 3.1. In 2006, a total of 431,942 passengers were carried on the inter-island services. That year SATA operated a total of 12,109 flights, in the process accumulating 6,151 flight hours and 7,722 block hours. Over the years, passenger traffic between the islands has grown consistently but at not a very high rate (Figure 3.2).

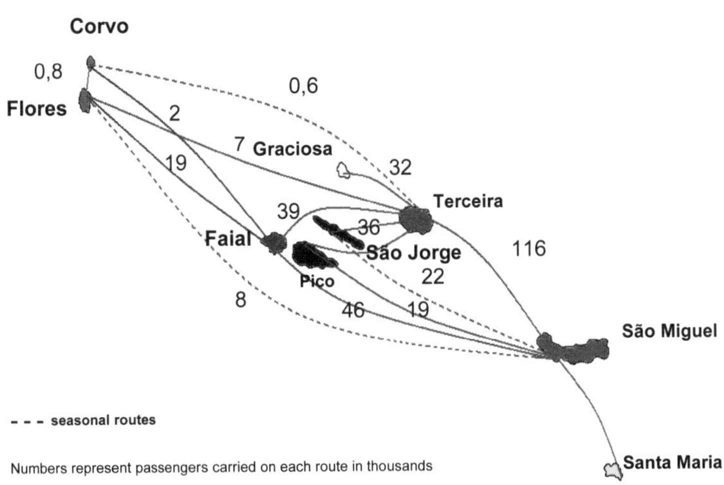

Figure 3.1 Inter-island annual traffic flow

1 For detailed information see the European Commission notice in the *Official Journal of the European Union* C49 of 28 February 2006, page 15 and C51 of 1 March 2006, page 7.

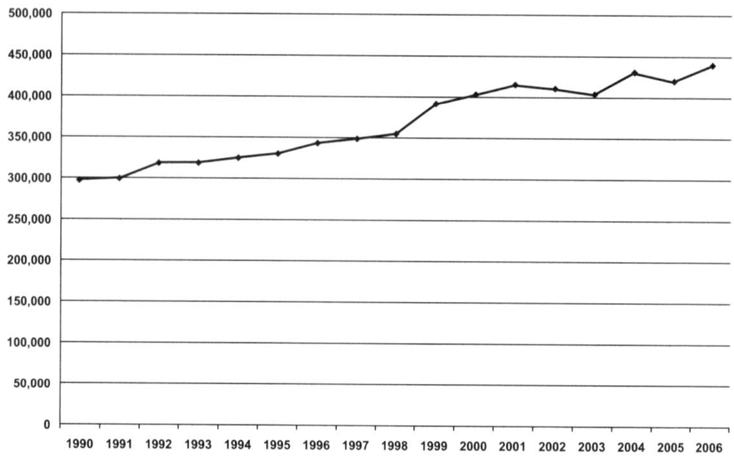

Figure 3.2 Inter-island passenger traffic evolution

It is possible to identify some of the factors that have influenced inter-island traffic development:

- In 1999 fares for residents travelling within the islands were reduced by 17 per cent.
- There was an increase in inbound traffic caused also by a fare reduction and increased capacity provided on the routes between the mainland and the islands.
- The PSO for the inter-island services established an increased number of frequencies and capacity.
- In 2005 there was a further reduction in inter-island fares, including lower fares for residents on routes between the less developed islands and the nearest business centre, and the introduction of a special lower fare for tourists travelling between the islands.
- Economic growth.

Mainland Azores Market Structure

The routes between the Portuguese mainland and the Azores, and also between Madeira and the Azores, were operated by TAP under a monopoly/concession until 1995. The first PSOs for these routes were established in 1995, followed by a public tender, but TAP was the only airline to bid. The PSOs on these routes were revised in 1998, introducing two main changes: each bid could now be for individual routes and not necessarily for all the routes as previously; the preferential fares for residents and the PEX fare were reduced significantly, by approximately

30 per cent and 20 per cent respectively. In the meantime SATA set up a subsidiary company, SATA Internacional, using jet aircraft, which started operation in 1998. This company was able to submit a tender for the mainland–Azores PSO routes for the period 1999–2001 and was awarded the PSO contract to operate the routes to Ponta Delgada (São Miguel).

In 2005, the PSOs were again revised. There is no longer a public tender; instead airlines must apply to operate the services on each PSO route for a period of one year in accordance with the conditions specified. In addition, the financial compensation is established on a fixed amount per flight coupon of resident passengers instead of an overall financial compensation as previously. TAP and SATA have been operating on a code-share basis since 2005. The last PSOs for the Azores–mainland air services were imposed on eight routes,[2] which are shown in Figure 3.3.

About 60 per cent of passengers travelling between the Azores and the Portuguese mainland are not residents of the islands. There has been a significant growth in this traffic, with the number of passengers travelling more than doubling between 1996 and 2006, as is shown in Table 3.8.

It is possible to identify some factors influencing traffic development between the Azores and the Portuguese mainland:

- the main growth happened after 1999 and was influenced by the new PSOs;
- increased number of frequencies and capacity;
- lower fares for residents and for inbound tourists;
- the entrance of a second operator (the regional airline) on the routes;
- increased hotel capacity in the islands;
- improved tourism promotion policy;
- economic growth.

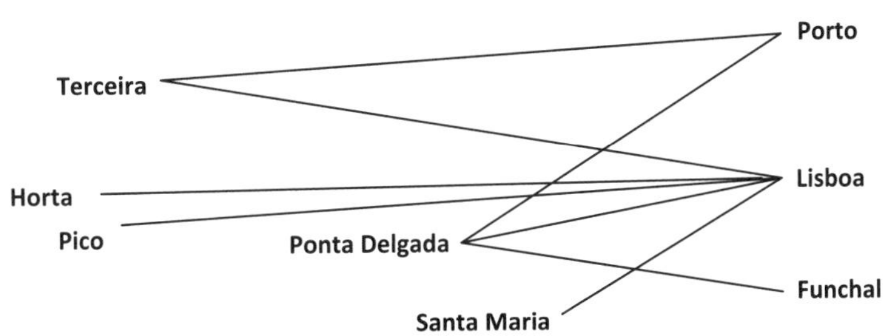

Figure 3.3 Current Azores–mainland Portugal PSO routes

2 *Official Journal of the European Union*, C104 of 25 April 2008, page 29.

Table 3.8 Growth in Azores–mainland Portugal passenger traffic

Route	1996 (TAP)	2006 (TAP + SATA)
Lisbon–Ponta Delgada	187,016	325,103
Lisbon–Terceira	91,886	172,858
Lisbon–Horta	34,239	84,418
Porto–Ponta Delgada	11,988	67,274
Funchal–Ponta Delgada	13,328	19,529
Lisbon–Pico	–	7,849
Lisbon–Santa Maria	–	4,970
Total	338,457	682,001

International Flights to the Azores

After airlines operating transatlantic services gave up stopping in Santa Maria, the only international flights to the Azores were operated by TAP to Boston, Massachusetts, with some seasonal charter flights from the US and Canada. Since 2000 SATA, the regional airline, has been operating weekly services to Boston and Toronto. A weekly scheduled flight from Ponta Delgada (São Miguel) to Frankfurt was also introduced. Since then, scheduled flights have also been launched to London (2005), Madrid (2005) and Amsterdam (2006). In 2000, the first seasonal charter flight between Ponta Delgada and Stockholm was operated. Nowadays a significant number of charter flights are operated by different charter airlines from Sweden, Finland, Norway and Denmark.

The most important international traffic to the Azores comes from Europe, the US and Canada. Passengers from the US and Canada are mainly travelling for visiting friends and relatives (VFR) purposes, the traffic being very seasonal and with little variation in levels of demand over time. By contrast, passengers travelling from Europe comprise predominantly tourists travelling on IT packages sold by traditional tour operators.

Passenger traffic disembarked at Azorean airports is growing consistently in all segments, inter-island, domestic and international, as is shown in Figures 3.4 and 3.5.

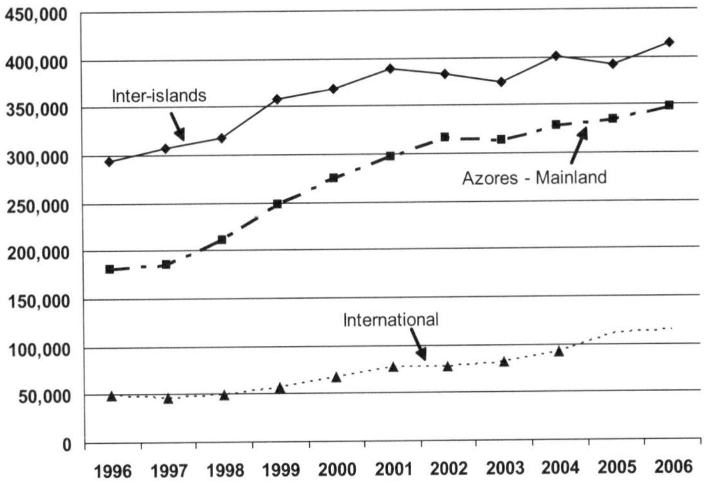

Figure 3.4 Passenger traffic at Azorean airports by type

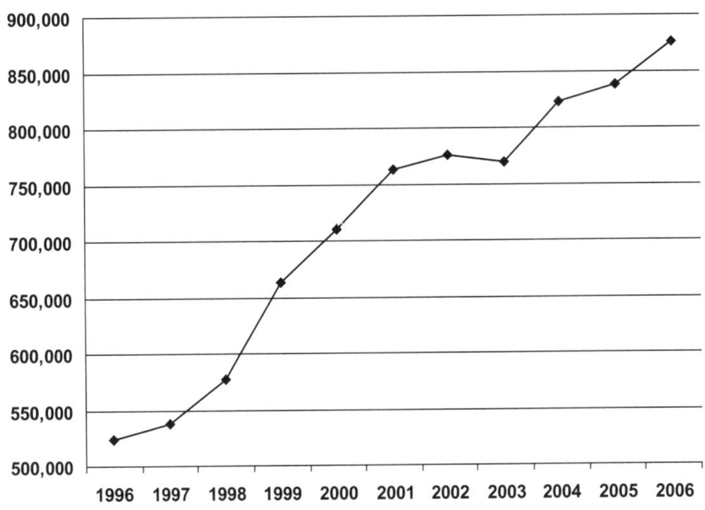

Figure 3.5 Total passengers disembarked at Azorean airports

Factors Influencing the Development of International Traffic to the Azores

There are several factors influencing the development of international traffic to the Azores, including:

- hotel capacity growth
- improved tourism promotion policy
- marketing incentives from airports
- marketing support to tour operators
- SATA's investment in developing new tourist routes to the Azores.

Growth of Hotel Capacity

Hotel capacity in the Azores has increased from 3,600 beds in 1996 to 8,200 beds in 2006, excluding guest houses and similar (see Figure 3.6). Fortunately, this is still far from an excessive hotel capacity, for example as on Madeira, which has more than 30,000 hotel beds.

Improved Tourism Promotion Policy

Naturally, the increased investment in infrastructure has resulted in the need to improve the tourism promotion policy by all the agents involved. The government and the tourism industry have founded an agency to support promotional activities. Although the agency's funding is relatively modest (€4 million in 2006), a significant number of promotional activities have been developed in the most important markets.

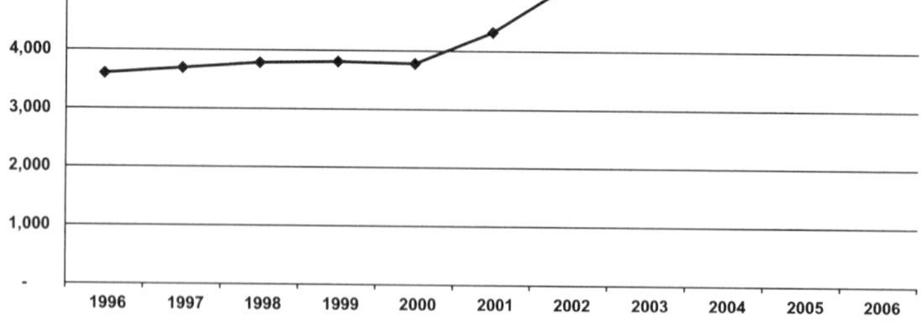

Figure 3.6 Hotel bed capacity in the Azores

More Aggressive Marketing Strategy by ANA

The Portuguese airport management company ANA realized that it could and should play an important role in route development. Airports now have a marketing strategy which involves the promotion of the region where the airport is located and the provision of financial incentive packages (discounts on airport taxes) for new international routes and for traffic development on existing international routes. Ponta Delgada Airport is also working in this direction, promoting and supporting new services.

Marketing Support for Tour Operators

Considering that new services to the Azores are essentially for inbound tourism, the Regional Government and Tourism Promotion Agency's contract with tour operators provides some financial support for the marketing activities needed to stimulate demand for the new flights. This support is more significant in the case of a tour operator assuming the risk of a charter operation.

SATA's Investment in Developing New Routes

As stated above, the new international routes serve mainly inbound tourism. Therefore, there is a higher risk involved in launching a new route for a destination where there is no other type of traffic type (e.g. business or VFR). SATA operated the first major charter series from Sweden to the Azores in 2000. Nowadays, charter flights from the Nordic countries to the Azores have increased greatly but are mainly operated by SAS and Finnair. Since 2000 SATA has launched scheduled routes from Ponta Delgada to Amsterdam, Frankfurt, London and Madrid. The strategy followed in respect of these services was to attract the largest possible number of small and medium-sized tour operators and to try to share operational risks with some of them. The most important of these routes are to London and Frankfurt. While SATA is the major player on the routes from North America to the Azores, it is not on the international routes to Europe.

Effects on Tourism

As a consequence of all these actions, there has been a significant improvement in tourism development. The number of bed nights spent in hotels has increased greatly. As Figure 3.7 shows, there has been growth both in the numbers of domestic and foreign guests staying in hotels. The number of domestic guests is higher than that of foreign guests, but the number of bed nights spent by foreigners surpasses the domestic market due to their length of stay, as shown in Figure 3.8 and Table 3.9. The proportion of hotel bed nights accounted for by domestic and foreign guests has changed dramatically since 1996, as may be seen in Table 3.9.

Nordic countries are responsible for 30 per cent of the total bed nights accounted for by foreigners in the Azores.

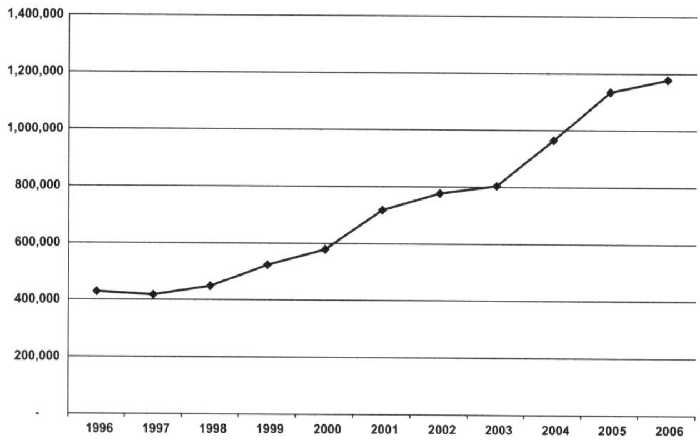

Figure 3.7 Bed nights demand in Azorean hotels

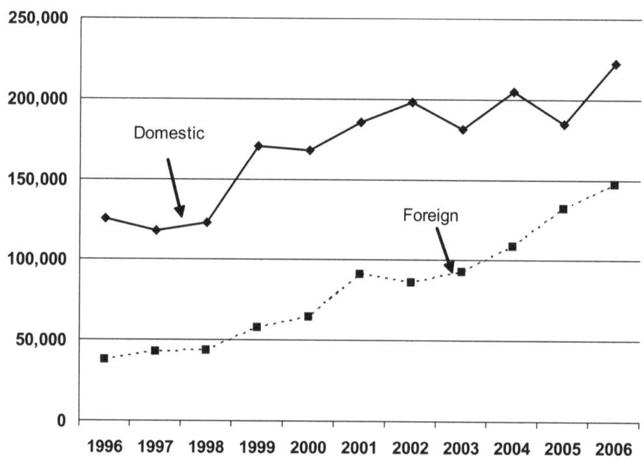

Figure 3.8 Domestic and foreign guests in Azorean hotels

Table 3.9 Hotel bed nights, domestic/non-resident share

	1996	2006
Domestic bed nights	72.7%	43.2%
Non-resident bed nights	27.3%	56.8%

Factors Constraining Route Development

Tourism is understood to be the key driver of economic growth in the Azores. It is also clear that air services are essential for tourism development. There are, however, a number of factors constraining the development of new routes, such as:

- The Azores is relatively unknown to the general public in Europe and North America.
- The Azores is a place for specific market niches rather than a mass destination.
- It is not easy to fill up aircraft every week with niche tourists.
- The remoteness of the islands and the small traffic volumes cause a price disadvantage when competing with the global international tourism industry.
- Except for specific markets (in particular, the Nordic market) or for seasonal operations it is not easy to sell the Azores in respect of charter flights.
- Major tour operators will not be interested in investing in a destination that will never produce a great number of passengers.
- So far it has not been possible to extend operations to the winter season on the European routes (except for the Nordic market).

Challenges

We could say that the attractiveness of the Azores results in part from thc fact that it is not a crowded destination. Aside from this, the islands are micro-systems that require the careful management of tourist flows in order to preserve their most valuable quality. There is, however, space in the Azores for some growth in tourism to occur. It is therefore necessary to continue to improve the promotion and marketing of the destination in order to achieve greater awareness in the main tourism generating markets and other more specific market niches.

Scheduled flights would appear to be better than charter flights to develop the market for small and medium-sized tour operators. The development and consolidation of scheduled flights on such thin routes represents a huge investment for a small airline. For example, a weekly frequency Ponta Delgada/Frankfurt/Ponta Delgada route operated year round with an A320 (165 seats) represents a cost exceeding €2 million. With an average occupancy of 75 per cent this represents only about 6,500 incoming tourists. An initial period with a realistic level of financial support is necessary to allow the launch of such routes and their viability over the medium term. A 'Route Development Fund' or an 'Essential Tourism Development Route Scheme' adapted from the schemes applied in other regions and taking into consideration the guidelines from the World Tourism

Organization and EU regulations would be a good start.[3] Naturally this would require further assessment and a deep involvement of the main beneficiaries of the new air services: government, the tourism promotion agency, the tourism industry, airports and airlines.

3 Nowadays in Portugal there are Route Development Funds in some regions such as the Algarve and Madeira. These funds result from partnerships between Tourism of Portugal, ANA (the airport management company) and regional promotion agencies. Each one contributes a certain percentage of the budget. The aim is to promote the launch of new routes with a tourism interest for the respective region. Perhaps one can say that in some circumstances the access conditions to the fund are too restrictive (in terms of minimum passenger numbers and frequencies) and that the support is a little limited (only €8–10 per passenger disembarked) to cover the costs of the launch of a new route.

Chapter 4
The Economic Impact of Air Transport in Remoter Regions

Jon Inge Lian

Introduction

Air transport has grown at a tremendous pace since 1950 due to its speed and its ability to shrink distances. Remote regions are per definition separated from central areas by distance and should thus be expected to be particularly dependent on air transport. Many countries have large areas that would be more or less isolated if not for air transport. This isolation could be due to long distances or barriers such as mountain ranges or sea access to islands. Air transport thus contributes to these regions by making them less isolated than without air transport. The benefits of a high-speed travel mode like air transport may be reflected in many ways: industries are better off, income levels are higher, population development is more favourable and the feeling of being remote is lower than if there were no air service.

In recent years there has been growing attention on the link between airports and economic development. In this chapter we will try to shed light on different impacts of air transport in remoter regions. We will not restrict the analysis to a very confined definition of economic impact, but will consider impacts in general, or impacts that at least have some bearing on economic development.

The chapter will first discuss the term *remoter regions*. Which aspects characterise remoter regions and what are the consequences of being remote? Then the term *economic impact* will be discussed. How can it be defined and measured and what are the methodological difficulties regarding the estimation of effects? Thereafter, examples of economic impacts will be given. The examples are drawn from a number of countries. Finally, the broader importance of regional airports with regard to population development and migration is described.

Remoter Regions – Distant and Small Markets

What characterises remote regions? The most obvious characteristic is that the regions are remote in the sense that they are distant from economic centres. A peripheral location implies long travel distances and very often a lack of other adequate transport services.

The other aspect of remoteness is that the market size is often small. If the market were large, the region would per definition not be distant from an economic centre. The implication of a small market size is twofold:

1. limited supply of local services both for businesses and the population;
2. limited supply of air transport.

The first feature was highlighted in Christaller's (1933) central place theory, stating that the range of services in small markets (which can be interpreted as remoter regions in our context) is small. There are few specialised higher-order services, for instance business services (ICT, banking, lawyers etc.), present in remoter regions. The implication of this is that there is a need for travel in order to acquire these services.

The second feature is due to the fact that a small market size can only support a limited frequency and level of service (type and size of aircraft). Further, there is usually no room for competition (between airlines). The consequence is that the regional service often will be a monopoly service with high prices if there is no public intervention.

Therefore, many countries have developed safeguards in order to offer an adequate air transport supply to smaller cities and remote areas. This has been done as a reaction to the geographical concentration of supply and hence changes in air fares, giving disadvantages to small communities under a competitive regime. In the US, the Essential Air Service Program (EAS) was established as part of the deregulation in 1978 (Reynolds-Feighan, 2001). In Europe, Public Service Obligation (PSO) was established from the beginning as part of the deregulation of air transport (Council Regulation 2407/92, article 4). The purpose of both these arrangements is to ensure a certain 'minimum' level of air service to remote areas if it is considered 'vital for the economic development of the region' (EU Regulation 2408/92). All these arrangements may be seen as an indication of the economic importance of regional air routes (Williams and Pagliari, 2004).

Air Transport and its Alternatives

The role of regional airports depends on *alternative opportunities to travel*. If there is fast rail service or highway that also connects the region, the airport will be of less importance. Further, authorities will probably adjust their policies to the transport situation. If there are other options, authorities will more likely be restrictive in developing and financing regional airports. Other considerations such as environmental aspects will then often be emphasised. However, if the situation is obvious, meaning that there is no suitable alternative to air transport, as in the case of Scotland and Norway, environmental considerations have to yield to economic and other arguments.

Air transport is especially convenient if there are physical or geographical barriers such as mountains or open sea that have to be overcome in order to reach islands. In these cases provision of other means of transport could be excessively expensive.

Another issue is access to congested airports for regional services. For instance, in the UK many regional authorities might want to increase the possibilities for regional services to connect at Heathrow to overseas services. National authorities need to protect the capacity at Heathrow and would favour alternative means of transport to Heathrow for the regions. The UK Department for Transport (DfT) has therefore suggested slot allocation mechanisms favouring major feeder routes, only securing access for 'peripheral regions' with total journey times to London Zone One by public surface transport of more than three hours (UK DfT, 2005).

Provision of air services has airports as a prerequisite. The airport should not be too distant, but should actually be located in the remote region it serves. Thus, both the impact of the air service and the impact of the airports in remoter regions should be the issues at stake here.

Types of Economic Impact

Intuitively, air transport is important for both the industry and the residents of a country. That is especially true if the alternatives to air transport are weak, due to long distances or thin markets. The importance of air transport becomes particularly clear when the air transport system fails, for instance due to strikes. However, it is not straightforward to measure the importance for society of a well-functioning air transport system, and it is often unthinkable to construct an alternative situation without air transport for comparison. The rapid growth in air transport in many economies may be seen as an indicator of the importance of air transport to economic growth. However, the direction of causation may be unclear. Rapid growth may at the same time reflect increasing travel opportunities for many people as a result of economic prosperity.

Air transport may be regarded as an economic activity in itself, where the resulting employment effects can be measured as direct, indirect and induced impacts derived from airport activities. These effects have been analysed in numerous studies, where the employment multipliers show a quite diversified picture of the link between airport activities and regional employment. Some of these variations may be due to differences in economic linkages in the airport regions, while others may result from the specific methods and data used.

If we concentrate on air transport as a contributor to economic activity more generally, the contribution can be characterised by four main types of effects, where the three first effects reflect air transport as an economic activity in itself, while the last effect stems from air transport as a catalytic agent inducing economic development in other industries (Figure 4.1).

Figure 4.1 Types of economic impacts

Type of impact	Description
1 Direct	Operation of airlines and airports (technical support and handling, catering, fuel, security, cleaning), commercial activities (shopping, restaurants, car rental, parking), land transport and air cargo.
2 Indirect	Sub supplies (goods and services) to direct activities (covered in the region)
3 Induced	Spending by employees in activity 1 and 2
4 Catalytic	a- Location impacts (firms and labour) b- Tourism and trade (demand side) c- Productivity and investment (supply side)

Direct Effects

Directs effects are defined as the activities required for providing the product, which in this case is air transport. The distinction between direct activities and indirect activities may be difficult to apply in practice. Usually, on-site activities, i.e. at the airport, are described as direct activities even if they are not air transport activities as such, for example other commercial activities at the airport.

The on-site principle, however, has its flaws, especially if indirect activity is calculated by the use of multipliers. At some airports, services are located at the airport itself, while at others the same services could be located outside the airport area, thus providing a smaller base for multiplier calculations.

The next question is what to measure and how to measure the extent of activity. Turnover figures might be difficult to acquire as the firms are complex, consisting of many units geographically spread and the quality of accounts for parts of firms may not be available. Employment figures might be much easier to acquire. Estimates of turnover and wages may be acquired later from structural business statistics from the Central Bureau of Statistics.

Indirect Effects

Indirect effects are defined as sub-supplies (goods and services) to direct activities covered in the surrounding region. A relevant airport region must therefore be defined.

Induced Effects

Induced effects are defined as impacts of income generated by direct and indirect effects, especially via private consumption (spending by employees on direct and indirect activities). Estimation of indirect and induced impacts is explained later.

Catalytic Effects

The catalytic effects are of particular interest because they indicate that air transport and airports may be catalytic agents, stimulating economic activity in industries that are not directly derived from or connected to air transport activities. In other words, catalytic effects indicate how access to air transport might serve as a prerequisite for economic activities. There are reasons to believe that these effects are significant, and that they may be even more important than the impacts derived from air transport activities by means of the more 'classical' multiplier effects.

When we initially stated that air transport is vital for industry and consumers, we were actually thinking about catalytic effects. These are a type of spill-over effects that capture the extent to which the growth in air transport boosts the performance of other industries (through tourism, trade, investment and productivity). It might well be the case that some important parts of the region's industry are less footloose because they are dependent on local skills and local industrial networks. If this kind of resource base becomes less productive without an airport in the vicinity, then there are productive catalytic effects present, and not only relocation effects that are often zero-sum games. Another issue is related to how economic players in remoter areas interact with their markets in buyer–supplier relationships. A priori, the probability of finding markets and collaborators is significantly higher for businesses in central areas. Hence, the air transport system may play an important role in exploiting the scale effects in both human capital and natural resources in remoter areas. However, there will certainly be large variations among regions depending on their existing resource base.

Catalytic effects can also include consumer welfare impacts as individuals benefit from increased possibility to travel. We distinguish between three types of catalytic effects. They are not fully commensurable as some of them may be regarded as measures (like location and trade) to achieve increased productivity (besides, trade may also be the result of increased productivity).

- location impacts (firms and labour)
- tourism and trade (demand side)
- productivity and investment (supply side).

When it comes to tourism and trade, there is an issue of positive and negative effects. The balance between trade and tourism may very well be negatively affected by increased air transport. The outward leakage of tourism may increase even if some regions benefit from increased incoming tourism. Besides, location effects may occur to the benefit of some regions at the expense of other regions. Positive cumulative growth effects in some regions may lead to negative backwash effects in others. This balance should be addressed carefully in order to identify the net national impacts. Positive airport impacts in remote regions may, however, contribute to national goals of developing such regions.

Cooper and Smith (2005) have estimated that EU residents travelling by air spend more as tourists outside the EU than visitors within the EU (approx. 0.3 per cent of GDP in 2003). The net impact was, however, positive for the ten new members of the EU, with spending by foreign visitors higher than spending by own residents abroad (net 0.4 per cent of GDP).

Cooper and Smith (2005) have also estimated effects on investments and overall productivity by controlling other drivers of economic growth. Regression analysis reveals a clear relation between air transport and investment development and the total factor productivity. If the use of air transport increases by 10 per cent (relative to GDP), that will boost business investments by 1.6 per cent in the long run. A long-term 1.6 per cent increase in capital stock per worker would in turn mean 0.6 per cent higher labour productivity. Correspondingly, a 10 per cent increase in air transport leads to 0.55 per cent higher total factor productivity in the long run. Effects seem to be greater in less developed countries than in developed economies.

The Air Transport Action Group has made similar calculations for world regions (ATAG, 2005). Total employment (including catalytic employment) is analysed in relation to the sum of direct, indirect and induced employment. The 'catalytic multiplier' varies from 1.15 in North America to 1.8 in Europe and 3.90 in Latin America. The less developed regions seem to benefit most from air transport. Transferred to a national perspective one could argue that remote regions, many of which are export oriented, would experience the largest catalytic effects.

The main objection against such econometric analysis is the causal direction. It is difficult to determine what comes first, growth in air transport or growth in the economy? Another consideration is city size. In order to create an attractive business environment, a certain size is necessary in order to ensure that there is broad range of qualifications present. An air route might be a necessary but not sufficient requirement in that respect. On the other hand, transport alternatives are less developed in remote regions, making air transport more vital for business development.

Studies on Multiplier Effects

As mentioned, employment effects can be measured as direct, indirect and induced impacts derived from airport activities. Employment multipliers vary from 1.3 to 4.1 (excluding the Valencia study with an extremely high multiplier). One could imagine that airports with a narrow definition of direct employment should have less employment per passenger (pax) and higher multipliers. However, a closer look at Table 4.1 shows that this is not the case. Still, the variations might be due to the different methods and data used, or they might also be due to differences in economic linkages in the airport regions.

Table 4.1 Employment multipliers at selected airports

Airport/year

Million pax	Multiplier (M): total/ direct employment	Direct employment per million pax	Total employment per million pax	
Paris 1996*	59.1	2.1	1,300	2,800
Paris 1991	44.8	2.3	1,600	3,700
Heathrow 1991	40.5	3.2	1,500	4,700
Gatwick 1996	24.3	1.7	1,000	1,700
Amsterdam 1997	31.0	1.5	1,600	2,400
Amsterdam 1993	21.3	2.0	1,700	3,400
Munich 1996	15.7	3.0	1,100	3,200
Düsseldorf 1997	15.5	2.2	800	1,700
Manchester 1993	13.4	1.6	1,500	2,300
Manchester 1991	10.8	3.6	1,000	3,600
Milan 1994	13.0	4.1	600	2,600
Copenhagen 1991	11.9	2.9	1,200	3,500
Oslo 1996	11.1	3.5	900	3,200
Barcelona 1994	10.7	2.0	500	900
Brussels 1993	10.0	1.5	2,000	3,000
Malaga 1993	6.3	3.0	400	1,200
Glasgow 1995	5.5	2.4	1,000	2,300
Birmingham 1994	4.9	1.5	1,000	1,500
Newcastle 1994	2.5	1.3	900	1,100
Valencia 1994	1.8	8.5	500	4,400
Cardiff 1997	1.2	1.3	1,600	2,100
Exeter 1994	0.2	1.7	2,800	4,900

*Charles de Gaulle (CDG) + Orly.

Sources: ACI (1998); Hakfoort et al. (2001); Strand (2002).

The direct employment at airports may vary according to airport size. Large airports might be more effective than small airports due to better exploitation of resources. On the other hand, large airports may have additional functions, for instance aircraft maintenance and repair, commercial activities etc. Looking only at the lower end of the scale (regional airports with mainly PSO services and some medium-sized airports in Norway), there seems to be clear tendency towards increased manpower per passenger as the airports get smaller. On average, there were 600 employees per million passengers in Norway in 2005 (Figure 4.2).

Direct employment could be observed by asking the airport operator or by sending questionnaires to firms involved at the airport. Indirect employment could also be recorded by questionnaires, but the amount of work and the degree of uncertainty in data collection and quality would be high. Induced employment, however, is difficult to observe directly. In most cases, some kind of economic model is used to estimate indirect and induced effects.

A Norwegian study (Lian et al., 2005) applied a regional economic model (Panda) to calculate the size of indirect and induced impacts of air transport sector activities in the hinterland of four airports. Panda is a single-region input-output model of a Keynesian type, which means that it is demand driven and that there is, by assumption, excess capacity in the regional economy. The model's main resource is the Regional National Accounts, which give data on intermediate deliveries from regional production sectors to each sub-activity.[1]

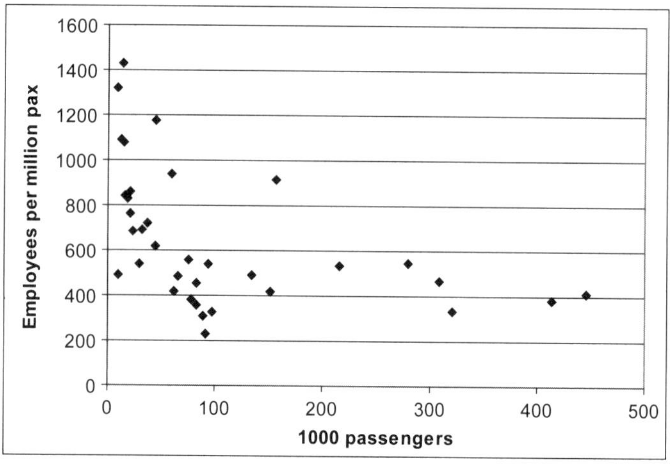

Figure 4.2 Employment per million passengers at Norwegian airports in 2005 by number of passengers

1 Fairly good representatives of air transport related activities in the model are needed. The following five sectors were used: government activities (Avinor, customs, police, airport security); land and air transport (airlines, fuel, transport to and from the airport); private services (parking, car rental, travel agencies, cleaning); hotels and restaurants (hotels, cafes, trade, catering) and post and telecom (goods handling and air freight).

Since the model is a relatively simple Keynesian input-output model, the (indirect and induced) impacts have to be interpreted as gross rather than net impacts. The study did not address the crowding-out effects of air transport, that is, the question of whether the resources allocated to air transport activities could be used more efficiently for other purposes. The study merely calculated the gross impacts of air transport activities. The model was run with and without the air transport activity.

The findings in Table 4.2 show that the multipliers vary according to the size and complexity of the hinterland. Larger airports have hinterlands with a broader spectrum of industries and services that capture most of the impacts within the region, while smaller regions experience a leakage of airport impacts to other regions (own coverage of services needed for air transport purposes is smaller).

Catalytic Effects – Discussion and Case Study

When measuring catalytic effects, measurement becomes even more difficult, especially since the direction of causation is not obvious. Do wealthy countries/regions get wealthy due to a well-developed air transport system? Or do wealthy regions have a greater demand for air transport and thus invest accordingly because they can afford it?

What is really needed are good control areas. The problem (for researchers) is the lack of comparable areas without air transport provision. Effects of infrastructure are long term and over the years nearly all regions have benefitted from improved infrastructure or transport services, making effects difficult to identify.

In addition, in studying regional airports there is a selection bias in that it is probably the most flourishing local communities that get airports in the first place. The catalytic effects are intuitively rather obvious. Without airports and air services, to do business or to run a society according to modern standards would be difficult in remote areas. However, the evidence of catalytic effects is very often anecdotal, based on interviews. One of the fallacies of interviewing is that strategic answers may exaggerate the importance of airports in order to get more public support for air transport.

Table 4.2 Multipliers at Norwegian airports (2005)

	Leknes	Molde	Bergen	Oslo
Multiplier, employment	1.37	1.26	1.61	1.91
Multiplier, man-years	1.33	1.34	1.54	1.96
Multiplier, production	1.54	1.49	1.71	1.99
Income multiplier	1.39	1.36	1.77	1.95
Million passengers	0.1	0.4	4.3	17.8

Therefore, the actual use of air services might shed some light on the importance of air transport in remoter regions. For instance on regional flights to Bodø and Tromsø in northern Norway, 20 per cent of air travel was for medical reasons. Patients were flown to regional and county hospitals for surgery and returned the same day. In other parts of Norway, oil and gas extraction is very important. In order to get the workers to the oil fields, air and helicopter services are vital.

The Norwegian national air travel survey of 2007 seems to indicate that the welfare benefit of airports is important in rural areas. There seem to be fewer business trips in rural areas than in cities. However, travel to/from work is important (Figure 4.3). In the case of Norway this implies travel to/from oil and gas activities.

A case study was performed in 2005 for Molde, a city with approximately 25,000 inhabitants, located on the north-western coast of Norway. The catchment area of the airport includes some 70,000 inhabitants. Employment is approximately 33,000 man-years. The industrial structure is dominated by the export of fish and advanced electro-mechanical products and engineering services. In addition, the city has a significant share of its employment (35 per cent) within the public sector, including a university college. The region had at the time of the survey the largest investment project in Europe within its borders, the Ormen Lange sub-sea natural gas project, with investments amounting to approx. €12 billion, including a pipeline to Easington in the UK.

The main objective of the survey was to study the share of the firms' activity level that was dependent on the existing air transport system. There were 78 respondents from industries in the catchment area, accounting for 22 per cent of employment in the area. The results from the study were:

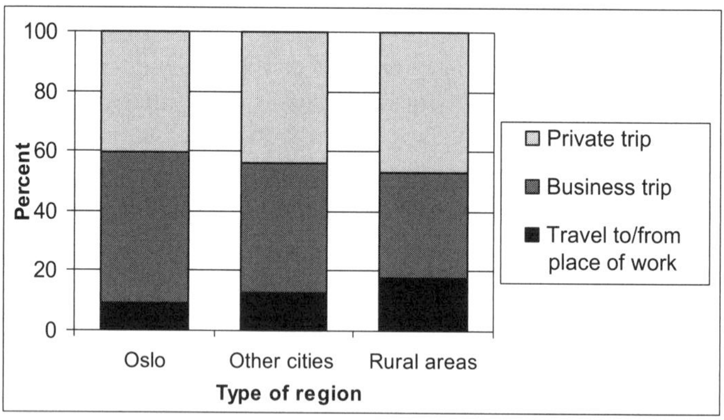

Figure 4.3 Air travel purpose by region of residence

- The average share of catalytic sales was 14 per cent for all respondents – 360,000 NOK per man-year.
- Respondents claimed 850 man-years as catalytic employment.
- Molde airport generates 143 man-years in total (direct+indirect+induced impacts).
- Compared to the average multiplier of 1.8 for European airports, catalytic employment appears to be very high. The number is, however, plausible when we consider the specific nature of the industries (oil and maritime).
- In total, catalytic sales amounted to 1.2 billion NOK, which may be 2–3 times higher if we generalise the region as a whole.

Other case studies indicate the importance of air travel for *tourism*. Remote areas need fast access for tourists in order to attract them. For some purposes, full-length runways able to accommodate larger jets are necessary otherwise the cost of air travel gets too expensive. Case studies in local communities in northern Norway show that high-yield adventure tourism is greatly dependent on air travel (Lian et al., 2007). Volume is, however, restricted by the level of service (high prices and small aircraft).

For Norway as a whole, incoming visitors by air stayed for 10 million nights, spending 13 billion NOK (€1.6 billion). However, Norwegian international air travellers spend about 50 per cent more abroad than do visitors to Norway. About 5 per cent of domestic travel is due to sports or cultural activities. Visiting passengers at Oslo airport spent €1.75 billion in the region, of which €1 billion was due to domestic visitors.

Air transport contributes to boost *productivity*, both in central and remote regions:

- Air transport makes cooperation and division of labour more efficient and influences industrial structure in the regions. Modern export-oriented industries are more dependent on air transport than other industries.
- Multinational companies and network industries locate in areas with good air transport provision (York Consulting Group, 2004). The same is probably the case at lower regional levels.

Another aspect of catalytic effects is the *value of reduced travel time*. Reduced travel time affects the productivity of the business traveller and increases the welfare of the leisure traveller. On the basis of Norwegian data, it is estimated that if 10 per cent of domestic air travel were transferred to the slower mode of car transport, it would amount to a yearly loss of 2 billion NOK for travellers due to the extra amount of time spent. Limited airport capacity, in a similar manner, gives rise to large economic costs due to the value of travel time saved if capacity is sufficient.

If climate-induced pricing instruments were introduced in Norway, resulting in a 10 per cent reduction in traffic volumes in all segments of the aviation market, the drop in activity would represent a loss of 600 million NOK (Lian et al., 2007). A 10 per cent reduction in air traffic gives a reduction in CO_2 emissions of 0.35 million tonnes. Given the contemporary estimated quota prices of approximately €25 per tonne CO_2, the value of reduced emissions amounts to 70 million NOK. Thus, given the current quota prices for CO_2 emissions, a climate tax would imply a huge loss to society. Future quota prices need to be nine times higher in order to match the societal loss.

However, there is no sense in a particular emission tax on aviation. Given that all production sectors and transport sectors are correctly priced, with prices equal to marginal costs, the most cost-effective way of reducing CO_2 emissions would be some sort of general economic measure applied to all sectors. If a certain result or level of emissions needs to be achieved, quotas would be a better measure than a general emission tax. At the moment it seems that air transport will be included in an EU CO_2 emission trading system. However, at the current level of quota prices, air travel will only be affected to a limited extent, since a quota price of €25 per tonne CO_2 equals 6 cents per litre of fuel and only €3 for a two-hour air trip per passenger. Other sectors besides air travel are more likely to be affected by increased quota prices.

Other Examples of Benefits to the Regions

In many countries, and also at the European level, regional development is supported by public policies. Some of the reasons may be cultural and historical in the sense that one wants to preserve traditional settlements. How successful is air transport as a policy to achieve regional development?

A Norwegian study has shown that there seems to be a clear relation between distance to airport on the one hand and population development and migration on the other (Lian, 2005). Regression analysis at the municipal level, controlling for other relevant variables, indicates that decreasing distance to an airport has a positive significant effect on population development (Figures 4.4 and 4.5). The columns are grouped by region. It is worth noting that population development varies significantly. The southern part of Norway has a positive development, while the northern part of Norway has mainly a negative development. Within each part of the country, distance to airport seems to matter.

Another relevant indicator was developed through a study made for the Norwegian Ministry of Regional Development. In a search for objective criteria for designating zones qualified for regional support mechanisms, travel time to the capital, Oslo, was introduced as one of several criteria. For the western and northern parts of Norway, travel time by air was measured. Norway is particularly dependent on air transport due to long distances (both domestic and to the European continent) and due to its sparse population and topography, making it expensive to build roads and railways. Several industries (for instance petroleum, tourism and the health sector) and regions (northern Norway and the west coast) are especially dependent on air transport.

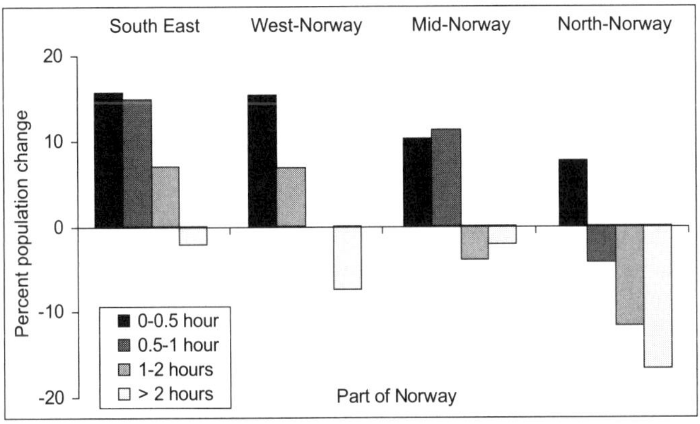

Figure 4.4 Population change (%) in municipalities 1990–2005 by distance to airport and region

Source: TØI report 807/2005.

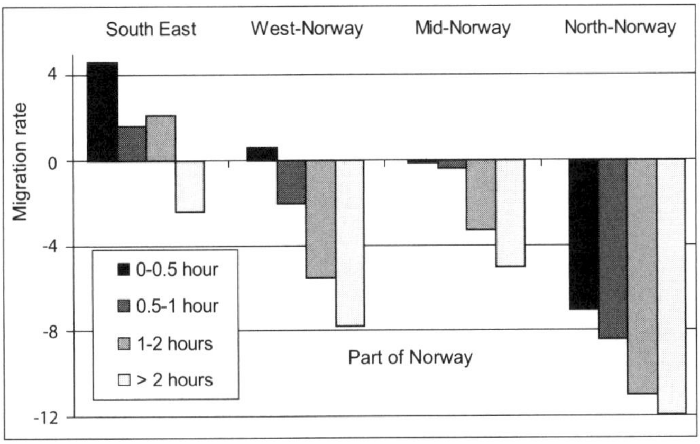

Figure 4.5 Net migration to municipalities 1994–2004 per 1,000 inhabitants in 2000 by distance to airport and region

Source: TØI report 807/2005.

In this study, it was shown that travel time was highly correlated to indicators of regional problems such as weak population development, low employment, low share of women of childbearing age, ageing population and low population density (Figure 4.6).[2] The direct interpretation of these findings is that remoteness matters and that air transport, by cutting down travel time, might counteract 'regional problems' created by distance (among other factors).

On the other hand, in spite of good airport coverage in Norway, these differences according to remoteness still seem to exist. Airport coverage is especially good in northern Norway and along the west coast, where two-thirds of the population have access to an airport within half an hour travel time due to the development of the short take-off and landing (STOL) airport system. Thus other factors must be of greater importance than air transport when it comes to explaining regional differences. Natural resources, industrial structure and the quality and size of the business environment are some other important factors. Once again air transport might be seen as a necessary but not sufficient requirement for business development in remote regions. The importance of air transport depends on the quality of transport alternatives. This importance might be increasing into the future due to shorter product cycles, increased communication levels and knowledge intensity of production and, consequently, a higher value of time.

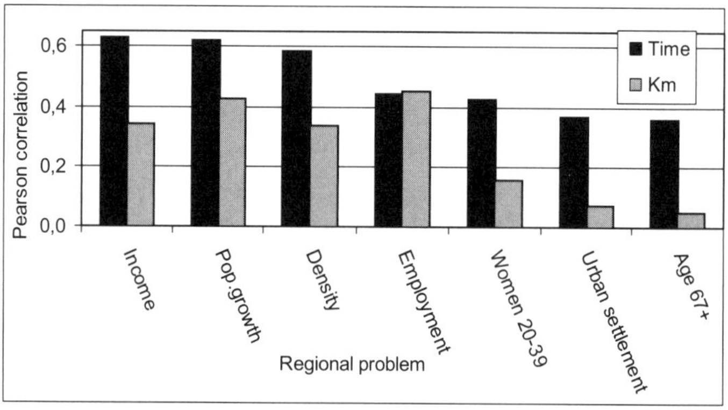

Figure 4.6 Correlation between travel time and distance to Oslo and indicators of 'regional problems' for Norwegian municipalities

2 All variables were defined with the 'worst' municipality as 100 and the 'best' as 0 (cut-off points at 10 per cent and 90 per cent of distribution).

References

ACI Europe (1998), *Creating Employment and Prosperity in Europe*. Brussels: ACI, 28s.

ATAG (2005), *The Economic and Social Benefits of Air Transport*. Air Transport Action Group.

Bråthen, S., Johansen, S. and Lian, J.I. (2006), *An Inquiry into the Link Between Air Transport and Employment in Norway*. Paper to the ETC conference 2006.

Cooper, A. and Smith, P. (2005), *The Economic Catalytic Effect of Air Transport in Europe*. Eurocontrol EEC/SEE/2005/004.

Hakfoort, J., Poot, T. and Rietveld, P. (2001), *The Regional Economic Impact of an Airport: The Case of Amsterdam Schipol Airport*. Regional Studies 35:7, 595–604.

Lian, J.I. et al. (2007), *The Economic and Environmental Impact of Air Transport in Norway*. Institute of Transport Economics, Oslo. Report no 921.

Lian, J.I., Bråthen, S., Johansen, S. and Strand, S. (2005), *The Economic Impact of Air Transport in Norway*. Institute of Transport Economics, Oslo. Report no 807.

Reynolds-Feighan, A. (2001), Traffic distribution in low-cost and full-service carrier networks in the US air transport market. *Journal of Air Transport Management* 7 (4), pp. 265–75.

Strand, S. (2002), *A survey of Economic and Geographic Impacts of Norway's Main Airport Gardermoen*. Institute of Transport Economics, Oslo. Report no 557.

UK DfT (2005), *Guidance on the Protection of Regional Air Access to London*. <http://www.dft.gov.uk/pgr/aviation/domestic/praa/annexbfullregulatoryimpactas2868>.

Williams G. and Pagliari R. (2004), *A Comparative Analysis of the Application And Use of Public Service Obligation in Air Transport within the EU*. Transport Policy 11, pp. 55–66.

York Consulting (2000), *Creating Employment and Prosperity in Europe: An Economic Impact Study Kit*. Brussels/London: ACI Europe/York Consulting, 18s.

York Consulting Group (2004), *The Social and Economic Benefits of Air Transport in Europe*. Brussels/London: ACI Europe/York Consulting.

Chapter 5
The Marketing of Small Regional Airports

Nigel Halpern

Introduction and Background

This chapter investigates the role of marketing as a means of developing air services at airports. The focus is on passenger services at smaller airports in Europe's remote regions (referred to hereafter as small regional airports). The chapter begins with some background on why marketing has become important as a means of developing air services at small regional airports. It then investigates how marketing can be used by airports to develop air services, providing examples from small regional airports.

Europe's remote regions have two main characteristics: they are sparsely populated and inaccessible to potential markets. They may also be characterised by permanent structural features such as an island location or a mountainous area (e.g. see Gloersen, 2005). Europe's remote regions generally have a good provision of airports. For instance, Table 5.1 lists a selection of remote regions in Europe. There are 96 airports, with an average of one airport per 75,379 inhabitants and 6,722 square kilometres. This compares to an EU19 average of one airport per 1,006,718 inhabitants and 9,412 square kilometres.

The European Commission classifies airports in Europe as being one of three types of connecting point: international, community or regional (see European Commission, 1996). Four of the airports listed in Table 5.1 fall into the international category (Keflavik, Faro, Palma de Mallorca and Inverness) and 12 of the airports fall into the community category (Tromsø, Bodø, Sumburgh, Kos, Rhodes, Ibiza, Menorca, Ajaccio, Olbia, Cagliari, Oulu and Luleå). The remaining 80 airports fall into the regional category. This means that they are relatively small airports with a total annual traffic volume of less than 900,000 passenger movements (of which less than 30 per cent are international).

It is the small regional airports that are of interest to this chapter. By serving a remote region, they are inaccessible to potential markets from where they can generate demand for inbound services. They also have a small catchment area from which to generate demand for outbound services. Their low traffic volume means that it may be difficult for them to convince airlines[1] of the need to establish new routes or expand existing services at the airport. It may also prove difficult to simply retain existing services.

---

1 In the case of charter airlines, it is likely to be a tour operator that is the main decision-maker regarding the development of air services. This chapter uses the term airlines when referring to both airlines and tour operators.

Table 5.1 Airport provision in a selection of remote regions in Europe

Region, country	Airports	Inhabitants per airport	Surface area (km²) per airport
Iceland	8	38,459	12,875
Faroe Islands	1	46,962	1,399
Northern Norway	28	16,508	4,034
Northern Sweden	10	50,947	16,530
Northern Finland	8	79,534	17,700
Highlands & Islands, Scotland	10	43,000	3,978
Border, Midland & Western, Ireland	5	232,180	6,650
Algarve, Portugal	1	421,528	4,996
Balearic Islands, Spain	3	338,135	1,664
Corsica, France	4	70,250	2,170
Sardinia, Italy	4	414,861	6,023
Dodecanese Islands, Greece	14	21,784	378
Total remote regions	96	75,379	6,722
EU19	402	1,006,718	9,412

Notes and data sources: Airport data is for 2004 and was extracted from the European Commission's Trans-European Transport Network Outline Plan. Population and surface area data is for 2007 and was extracted from Eurostat's database for regional statistics. EU19 data is for 2007 for the 19 pre-2004 accession member states of the European Economic Area and was extracted from Eurostat's database for regional statistics.

Despite the challenges to developing air services, small regional airports have a number of opportunities that they can exploit. They are likely to be uncongested and have plenty of spare capacity. The inaccessibility of the region means that air transport may be the only viable way for people to travel to and from the region for business or leisure purposes. In addition, remote regions are often blessed with natural resources such as extreme climatic conditions, islands, beaches, mountains, wilderness landscapes and wildlife. These resources can be exploited for the development of inbound tourism, with air travel being the fastest and most efficient means of travelling to and from the destination.

European air transport markets were deregulated by 1997 and this has created further challenges and opportunities for small regional airports. Prior to deregulation, domestic services to small regional airports were somewhat protected because, in exchange for monopoly rights on dense and profitable routes, governments would often require national airlines or their subsidiaries to serve lightly populated and unprofitable routes such as those serving small regional airports (Williams, 2002). International services were also dictated by governments under bilateral air service agreements (see Doganis, 2006); but such agreements would rarely involve small regional airports where the focus was largely on providing domestic services that linked the small and isolated communities to the main transport networks, thus improving the accessibility of their region (Reynolds-Feighan, 1995).

The only scope for airports to compete was for charter airlines that fell outside of the regulatory regime. Charter airlines helped shape international networks between regional airports, primarily serving package holiday markets flowing from northern to southern European destinations (Papatheodorou, 2002). Many of the larger airports in Europe's remote regions, such as Palma de Mallorca, Rhodes and Ibiza, benefited from such markets. However, the decisions of airlines were still largely driven by governments that were willing to invest in the necessary regional infrastructure. For instance, the boom in Spain's package holiday industry was influenced by the state under Franco's centralist policy (Baidal, 2004).

Deregulation has caused a strong reduction in cross-subsidisation of loss-making domestic services. It has also resulted in the elimination of bilateral air service agreements for intra-European services. A system of regulation does still exist in Europe. The system is implemented through the Public Service Obligation (PSO) programme, which supports the provision of lifeline services at a number of small regional airports. PSOs work well on particularly thin routes where there is 'no other option'. However, some PSOs might be seen as a constraint to future development because they do not allow competition from non-contract airlines and can result in the provision of minimal services with small aircraft and high fares (e.g. see Halpern and Niskala, 2008). As a result, airports served only by PSOs often aspire to develop other commercial air services not in direct competition with the PSO services.

Deregulation has resulted in increased competition between airlines, even if mergers, acquisitions and also low traffic have left some markets as monopolised. In other markets, competition has been encouraged further by the introduction of low-cost airlines. A number of low-cost airlines now serve small regional airports (e.g. Norwegian serves Kirkenes, Bardufoss and Harstad/Narvik in northern Norway; Blue 1 serves Rovaniemi and Kittilä in northern Finland; Iceland Express serves Akureyri in Iceland; Ryanair serves Alghero on the island of Sardinia; and easyJet serves Mykonos on the island of Mykonos in Greece). Deregulation, along with more general factors such as economic growth, has resulted in increased demand for air travel in Europe, providing opportunities for more airports to develop air services. For instance, there were 56,382 additional flight movements from commercial passenger airlines in Europe between January 2006 and December 2007 compared to between January 2005 and December 2006; 64.8 per cent of the additional flight movements were made by low-cost airlines, 29.6 per cent by traditional mainline or regional airlines and 5.6 per cent by charter airlines (Eurocontrol, 2008).

Deregulation has meant that airlines in Europe have much more freedom in choosing which airports they serve. This offers both threats and opportunities for small regional airports because, despite being threatened by the loss or reduction of services that are not commercially viable, they have the opportunity to attract new routes or expand existing services that were previously constrained by regulatory barriers. However, in order to exploit the opportunities, small regional airports need to market themselves to airlines. This is because the decisions of airlines in Europe are now driven largely by market forces rather than public service considerations. Marketing can also allow small regional airports to overcome the challenges of serving a remote region and exploit any opportunities.

Marketing

Marketing theory uses the concept of the marketing mix, a tactical toolkit that can be used by companies to exploit market trends and to position their product/service in the marketplace. The individual elements of the marketing mix are widely known as the four Ps that were first introduced by McCarthy (1960):

1. Product – the product/service that is produced.
2. Promotion – the methods used to market the product/service.
3. Price – the amount that is charged for the product/service.
4. Place – the distribution channels used to sell the product/service.

The four Ps have been reinvented on a number of occasions. For instance, the four Ps have been extended to the seven Ps in order to include three service-orientated elements: physical evidence, processes and people (see Brassington and Pettit, 2007). More recently, Lauterborn (1990) has suggested that the four Ps should be replaced with the four Cs in order to reflect a more customer-orientated philosophy, where product is replaced by customer needs and wants, promotion is replaced by communication, price is replaced by cost to the user and place is replaced by convenience. Despite the range of philosophies, the fundamental principles are the same and the original four Ps provide a sufficient framework for this chapter.

Product

According to Graham (2003), the airport product consists of a supply of tangible and intangible services that can be divided into three main elements:

1. The core product which is the benefit sought (i.e. the ability to land and take off).
2. The physical product that delivers the benefit sought (i.e. tangible infrastructure such as terminal and runway capacity or intangible services such as ground handling and security).
3. The augmented product that provides additional benefits to the core and physical product (i.e. service levels).

Airports deliver the core product as a matter of cause. They must then ensure that the physical product meets the needs of their target markets, although, to some extent, the physical product actually determines the types of market that can be targeted. Only airports with adequate runway dimensions, clearance for international flights and appropriate terminal capacity can realistically target airlines that are able to operate dense intra-European routes with large aircraft. Airports with smaller runways, domestic clearance only and limited terminal capacity will need to target airlines that are able to operate thin routes such as domestic hub connections with small aircraft.

The harsh operating conditions that are typically associated with remote parts of northern Europe (i.e. frequent adverse weather and permanent obstacles such as mountains) may provide further constraints to airports that are not equipped with modern landing systems and accurate real-time weather monitoring, both of which have the capacity to improve airport safety and the reliability of flight operations. The provision of adequate ground handling and security services can also improve airport safety and the reliability of flight operations.

Competition between airports that have similar levels of infrastructure and services will typically take place at the augmented level. Cost reduction has become a key feature in the airline industry since deregulation as lower costs allow airlines to be more competitive with the fares that they offer to passengers. Airports can exploit this need at the augmented level by providing service levels that allow airlines to maximise the utilisation of aircraft. For instance, this might be achieved by ensuring minimal delays and a fast turnaround of aircraft, by providing multi-functional and flexible staffing, or by having long airport operating hours.

Rovaniemi Airport provides a good example of the role that the airport product plays in meeting the needs of target markets. The airport is located in northern Finland and serves the winter charter market in Lapland. The airport has experienced strong growth in demand from international charter airlines since they first started flying to the airport in 1984. In order to meet increased demand, the airport has had to continually modify elements of its physical product. Figure 5.1 illustrates the growth in international passengers at Rovaniemi Airport and it can be seen that growth was particularly strong after each of the improvements in infrastructure had been completed (i.e. 1993 to 1995, 1996 to 1999 and 2000 to 2004).

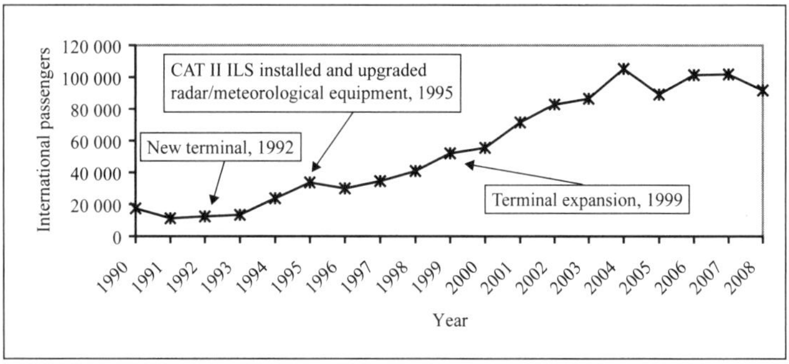

Figure 5.1 International passengers at Rovaniemi Airport, 1990–2008

Notes: CAT II ILS is a second category instrument landing system that provides precise guidance to an aircraft for approach and landing with a decision height greater than 30 metres but lower than 60 metres above touchdown, and a runway visual range of no less than 350 metres.

Source: Finavia (2009).

Meeting needs at the augmented level is also a priority for Rovaniemi Airport and for other airports in northern Finland that serve the winter charter market in Lapland. The following quote emphasises the focus on ensuring minimal delays and providing flexible staffing:

> During peak periods, parking aircraft seamlessly and conducting thousands of passengers through terminals designed for small numbers pose an annual challenge for the [Lapland's] airports. For all this to work we have to have seamless to-the-minute teamwork between the airport's staff, the airlines, the Border Guard service, Finnish Customs, the tour organisers, the safari and handling companies and others. Terminal services, air traffic controllers and airport maintenance teams need to be especially flexible in order to get the aeroplanes to land and take off on time. (Finnish Civil Aviation Administration, 2004, p. 27)

Associated with the airport product is the idea of the airport brand. This can be represented by a name, logo or design, or by a particular style of signing, merchandising or advertising (Graham, 2003). Branding creates distinctiveness and adds tangible cues to what is essentially an intangible service. Branding can also promote recognition, preference and loyalty amongst target markets. The brand that is developed may be based upon a number of features such as:

- The region that is served (e.g. Ireland West Knock Airport, Girona Costa Brava Airport);

- Natural attractions (e.g. Lakselv Banak North Cape Airport, Annecy Haute-Savoie Airport Mont Blanc);
- Man-made attractions (e.g. Bardufoss Snowman International Airport, Rovaniemi Santa Claus Airport);
- Aspects of historical importance such as famous people (e.g. Kavala International Airport Alexander the Great, Federico García Lorca Granada-Jaén Airport);
- The size and scope of services available, especially international services (e.g. Perpignan-Rivesaltes International Airport, Evenes Lofoten International Airport).

Whether branding provides airports with a competitive advantage or not is subject to debate (Graham, 2003). Indeed, branding can have a negative impact by being too distinctive. This can encourage aspects such as seasonality. To help facilitate the winter charter market in Lapland, Rovaniemi Airport was branded. It bears the Official Airport of Santa Claus trademark within the European Union and several other countries. As can be seen from Figure 5.1, the initiative has coincided with growth in demand from international passengers. However, demand is highly seasonal (see Figure 5.2): 90 per cent of the 91,697 international passengers in 2008 were served during January, February, March and December. Almost 60 per cent were served during December alone (Finavia, 2009).

Seasonal demand results in an inefficient use of infrastructure and services. Investment may be needed in order to facilitate growth, maintain service levels and reduce the risk of losing demand to competing airports. However, it may be difficult to justify investment for a peak period that lasts for just a few months of the year. This poses a challenge for any small regional airport that is focused on developing seasonal high-density air services.

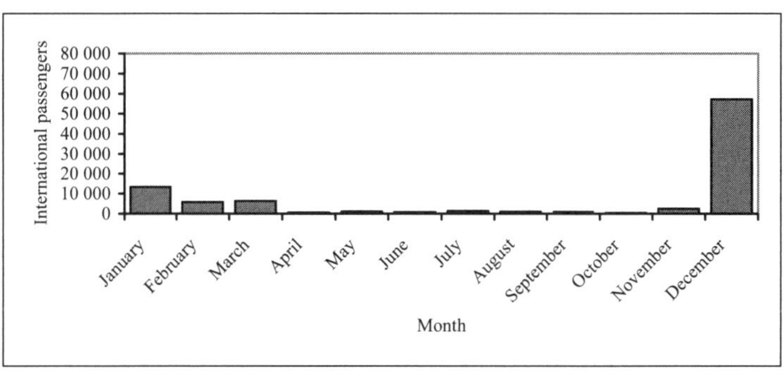

Figure 5.2 International passengers at Rovaniemi Airport, 2008

Source: Finavia (2009).

Promotion

Marketing theory typically refers to the promotional mix: five main tools that can be used by companies to persuasively communicate with target markets:

1. Advertising – non-personal means of mass communication
2. Direct marketing – the development of personal relationships with individual customers through the use of direct mailing, telephone selling or electronic media. Direct marketing may be considered as a form of advertising
3. Personal selling – personally tailored messages delivered face to face
4. Sales promotion – short-term incentives to generate sales
5. Public relations – building good relations with interest groups such as the media, shareholders and trade unions.

Public relations are important for small regional airports but this, along with sales promotion, is used less extensively than the other tools as a means of developing air services.

Advertising is used extensively by small regional airports to create awareness and communicate certain messages to target markets. It can be carried out in a range of ways such as producing general publicity information (i.e. advertisements, information leaflets, stickers and T-shirts), advertising in a range of media (i.e. newspapers, radio, television and electronic media) and attending relevant conferences and exhibitions (i.e. representation at travel trade exhibitions, which is particularly important when targeting charter airlines rather than scheduled airlines – the latter increasingly reduces links with the travel trade in order to reduce distribution costs). Catchy slogans may be used to promote the airport. For instance:

- Grenoble Isère Airport – the low-cost airport of the Rhone-Alps region
- Chambéry Savoie Airport – an airport of proxim. A door on the French Alps
- Kemi-Tornio Airport – for golf in the midnight sun
- Enontekiö Airport – the airport in the mountains
- Lycksele Airport – the centre of southern Lapland
- Hemavan Airport – fly into adventure.

Pajala-Ylläs Airport is located in northern Sweden and international charter airlines are one of the main target markets of the airport. The airport has produced a range of publicity information about the infrastructure and services available and tourism activities in the region. Various giveaways have been produced such as airport stickers and T-shirts. The airport has also produced a double-sided card (see Figure 5.3). The front of the card emphasises the smooth and low-cost landings available at the airport and its location in the tourist destination of Lapland. The winter wonderland depicted by the picture further emphasises the tourism potential

Figure 5.3 Publicity information for Pajala-Ylläs Airport

of the region. The back of the card illustrates the close proximity of the airport to the mountain resort of Ylläs in Finland. It also emphasises the infrastructure and services available at the airport and provides the contact details of the airport manager. The airport has also produced advertisements and an airport video and has been represented at a number of aviation conferences and travel trade exhibitions. Advertising often uses the slogan 'your next destination in Lapland'.

The problem with advertising is that it only communicates general messages to a general audience. It rarely says anything about the catchment area or potential for demand. It can also be very costly. This is especially the case for airports seeking to advertise in airline or travel trade magazines such as *Airline Business*, where an advertisement costs €11,285.[2]

Methods of direct marketing can be used to target a specific audience but the messages are still likely to be fairly general. Instead, airports tend to use personal selling as a means of communicating directly with target markets, normally through what has become known as an airline presentation. This provides specific airlines with a detailed analysis of the potential for a particular route. The analysis can then be supplemented with additional information on the catchment area and potential for demand and on the infrastructure and services available at the airport and in the region in general. Airline presentations have become standard practice in the aviation industry. The analysis therefore needs to be based on sound market research in order to make a convincing case. The analysis also needs to be realistic. Exaggerated claims about route potential will only lose the trust of target markets and may affect an airport's ability to develop future air services. Chambéry-Savoie

2 The price quoted is for a one-page colour advertisement in 2009.

Airport, which is located in French alpine region of Savoie, places a great deal of emphasis on market research, stating that:

> Our marketing department has obtained all the latest databases for traffic and holiday travel available on the market. Combined with our powerful modelling and reporting tools, this has enabled us to study the potential, in real time, of any service departing Chambéry, compared with other destinations in France. We can study the potential of any route to or from other airports in France and compare it with the profitability it could expect to achieve operating out of our airport. A sophisticated QSI [Quality of Service Index] model, that has already been tested and proven, has also been developed in-house. The marketing department will be happy to give existing and prospective customers the benefit of all its studies, models and tools. (Chambéry-Savoie Airport, 2008)

One development that has supported, and increased, the use of personal selling is Routes – The World Route Development Forum. Routes is a type of speed dating for airports and airlines as it provides networking opportunities through one-to-one meetings. During the meetings, airports typically use market research to demonstrate opportunities for new routes or expanding services at the airport. Routes – The World Route Development Forum was founded in 1995 and is held each year along with region-specific events such as Routes Regional Americas, Routes Regional Asia, Routes Regional Europe and Routes Regional Africa. Members of Airports Council International-North America launched a similar event called the JumpStart Air Service Development Program. This was launched in 1997 and is held each year in North America.

Although airports are in a strong position to research and promote opportunities for the development of air services at their airport, they must accept that they are a derived demand and that only on very rare occasions will a passenger travel to an airport in order to visit that airport. This means that airports need to be aware of the needs of local stakeholders and the wider objectives of their region. Collaboration with local stakeholders such as tourism and regional development agencies is important and may subsequently enable airports and local stakeholders to pool resources, develop an integrated approach to regional development and provide target markets with increased support and a wider overview of the area and its potential. Collaborations of this nature are particularly important for small regional airports where resources to conduct their own research and promotional activities may be limited. Small regional airports or airport groups that have been a part of a regional collaboration include:

- Hemavan Airport – a member of Tärnafjällen Incoming
- Lakselv Banak North Cape Airport – a member of North Cape Airport Services
- Molde Airport Årø – a member of Molde Airport Development

- Highlands and Islands Airports Limited – a member of Highlands Loch Ness marketing group.

Collaborations have been taken a step further in some regions through the establishment of Route Development Funds (RDFs). RDFs were initially established in parts of Scotland and Northern Ireland (e.g. see Kealey, 2004; Eden, 2005) and are typically based on a partnership between the public sector, airports and airlines. RDFs are typically aimed at encouraging new air services that promote business links and inbound tourism so as to benefit a particular region. The funds enable precise economic evaluations of particular routes and reduce the start-up risk to airlines agreeing to operate such routes by providing them with financial support. However, RDFs in Scotland and Northern Ireland have produced mixed results. Pagliari (2005) points out that only nine of the 17 routes established with the support of Scottish RDFs between March 2003 and May 2004 survived. In addition, although airports can lobby for such schemes to be established, they tend to be driven by national or regional governments so the extent to which airports can establish such schemes is fairly limited.

Price

When airports promote opportunities for the development of air services, airlines will expect to find out how much it is going to cost them. Pricing is therefore a key principle. It represents the sum of the values that the airline exchanges for the benefits of using the airport and ultimately determines whether or not an airline commits to developing air services at an airport.

Airports typically encounter a number of internal and external forces that influence their pricing decisions. Internal forces include the marketing objectives and the costs of the airport, whilst external forces include the nature of the market and demand for air services, the level of competition for air services and issues relating to the wider business environment.

The low volumes of traffic at small regional airports mean that their marketing objective is often to develop air services. This compares to busy, capacity-constrained airports where the marketing objective may be more focused on public relations activities. Cost issues at small regional airports also support the need to develop air services. This is because many small regional airports are loss making and have very high fixed costs in terms of labour, maintenance and utilities. The marginal cost of accommodating additional air services is likely to be very low, whilst the revenue gained from increased passenger throughput (i.e. from car parking, retail and catering) can be significant. The social and economic impact of airports on their surrounding area is discussed in numerous studies (e.g. see ATAG, 2008; ACI-Europe, 2004). This means that local stakeholders such as tourism and regional development agencies may support the need to develop air services at their airport and encourage their airport to develop air services.

External forces also influence the price decisions of small regional airports. Their low traffic volumes mean that the nature of the market and demand for air services are likely to be fairly limited. This implies a fair amount of risk for an airline when establishing a new route and there is an incentive for airports to adopt a pricing policy that shares in such risk. The level of competition between airports is another external force that affects price. Competition is likely to be particularly high between airports seeking to compete for inbound services such as those supporting the development of inbound tourism. This is because airlines may have a number of airports and destinations that they can choose from.

Issues relating to the wider business environment also influence the pricing decisions of small regional airports. For instance, current fluctuating fuel costs mean that airlines are seeking to save costs by reducing capacity or dropping routes, especially at airports where costs are particularly high. These airports may need to reduce prices or offer incentives in order to maintain existing services, let alone attract new routes or expand existing services.

The greatest individual influence on airport pricing is the extent to which airports have control over their pricing. According to Halpern and Pagliari (2007), most small regional airports are publicly owned. The prices that they charge are typically set and controlled by regulation so the airports have little control over pricing. In addition, many small regional airports are operated as part of a national or regional airport system where prices are the same for all airports in the system. Both of these factors affect the ability of small regional airports to compete on price. For instance, 46 airports in Norway are operated by Avinor as part of a national airport system. Airport user charges are levied in the same way for all airports in the system and are approved annually by the Norwegian Ministry of Transport and Communications.

Airports are not always fully responsible for their services provided and this can also affect their ability to compete on price. For instance, individual airports are not always responsible for the provision of ground handling services or any government taxes or en-route charges that might be levied for using the airport. This means that they will not be able to fully control the overall cost of the airport usage. However, they may be able to reduce certain aspects of the overall cost by offering incentives that complement any published airport charges.

Price incentives have become commonplace at many small regional airports, especially at airports seeking to attract new routes. Incentives typically offer discounted airport user charges that diminish over a set period of time and as the service becomes more established. They demonstrate that the airport is willing to share in the high start-up costs and risk associated with establishing a new route. Waterford Airport is a small regional airport located in the south east of Ireland. The airport user charges for 2008 included a passenger handling charge of €10 per departing passenger and a security charge of €2.50 per departing passenger. Where an airline enters into a three-year agreement for a new route, the charges are discounted (see Table 5.2).

Many small regional airports offer other types of incentive such as based aircraft incentives and growth incentives. The former encourages airlines to base their aircraft at the airport. This subsequently encourages a greater use of airport infrastructure and services. It also encourages loyalty as an airline may be more committed to develop air services to and from a base airport. The latter encourages airlines to increase passenger demand for new or existing routes (i.e. by improving frequency or capacity, timing of flights or promotional activities). The airport user charges for Waterford Airport in 2008 included a growth incentive scheme for new routes by offering airlines a rebate on the discounted passenger handling charge (see Table 5.3).

Table 5.2 Price incentives for new routes at Waterford Airport, 2008

Year	Passenger handling charge discount	Security charge discount
1	50% to €5.00 per departing passenger	40% to €1.50 per departing passenger
2	35% to €6.50 per departing passenger	30% to €1.75 per departing passenger
3	25% to €7.50 per departing passenger	20% to €2.00 per departing passenger

Source: Waterford Airport (2008).

Table 5.3 Growth incentives for new routes at Waterford Airport, 2008

Year	Rebate
1	50% of the €5 discounted handling charge applying for the year multiplied by the total number of departing passengers for the period less 60,000
2	50% of the €6.50 discounted handling charge applying for the year multiplied by the total number of departing passengers for the period less 67,500
3	50% of the €7.50 discounted handling charge applying for the year multiplied by the total number of departing passengers for the period less 75,000

Source: Waterford Airport (2008).

Another way in which airports are known to offer price incentives is through providing marketing support for joint advertising and promotional campaigns. Airports have been known to develop advertising and promotional campaigns of their own to generate demand for services at their airport. For instance, Ireland West Airport Knock, which is located in the west of Ireland, launched a series of television adverts in the UK in August 2007 to attract more tourists to the region during the autumn and winter months. However, airports are a derived demand. This means that instead of conducting their own advertising and promotional campaigns to passengers, it may be more effective for them to support campaigns via intermediaries such as airlines (including tour operators), travel agents or travel planning portals. Intermediaries have a strong level of brand recognition amongst passengers and may be able to penetrate markets more effectively than airports through aggressive marketing campaigns for specific routes.

Molde Airport Årø, which is located in the west of Norway, has supported a range of joint advertising and promotional campaigns to promote its outbound charter services to Gran Canaria for the 2008/2009 winter season. Figure 5.4 provides an example of an advertisement that has featured in a range of places such as local shops, newspapers and at the airport itself. The advertisement is in Norwegian. It advertises the offer of direct flights with the tour operator Star Tour from Molde to sun and summer in Gran Canaria between 15 October 2008 and 8 April 2009. The advertisement includes a selection of photos of the Star Tour products in Gran Canaria, including BlueVillage, BlueSelection, BlueCouples, The Bear Club (bamseklubben) and Super Kids. Molde Airport Årø and a local airport development group called Molde Lufthavnutvikling advertise the availability of discounted car parking (100 Norwegian kroner per week[3]) and duty-free shopping for passengers using outbound charter services to Gran Canaria. The advertisement states that the holiday can be booked at local travel agents (HRG or Bennett Ferie) or online at www.startour.no. The logos of different stakeholders feature on the advertisement, emphasising its joint nature. Molde Airport Årø has also supported the use of advertisements in the airport, at the local ferry terminal and on local buses (in digital, placard and printed format); advertisements on local television; displays in the airport and at local shopping centres; direct mailing; and newsletters. One advertisement consists of a highly visible banner that is 3 metres high and 5 metres wide and is displayed on the outside of the airport terminal building. In 2008, Molde Airport Årø contributed approximately 350,000 Norwegian kroner[4] to the marketing of its international charter services (including the service to Gran Canaria for the 2008/2009 winter season), duty-free shopping and car parking.

The Norwegian airport operator Avinor offers airlines the opportunity to apply for marketing support to promote new routes or for other marketing initiatives at their airports. The marketing support may contribute towards:

- special activities at the gate;
- special events for invited persons such as airline representatives or the press;
- temporary dedicated check-in areas;
- marketing campaigns directed at travellers to raise awareness;
- the production of general publicity information or advertising in a range of media, including newsletters, reports on the radio, press conferences or articles in magazines;
- establishing internet links between airports and airlines or destinations.

---

3 As of 17 December 2008, 100 Norwegian kroner was equivalent to €10.46.

4 As of 17 December 2008, 350,000 Norwegian kroner was equivalent to €36,665.75.

Figure 5.4 Joint advertising and promotional campaign, Molde Airport Årø

The amount awarded by Avinor is a one-off payment that can cover up to 50 per cent of the external costs of the marketing activity. The external costs are defined as the direct expenses involved, excluding personnel costs or other direct costs related to the funding partners or owners of the project. In addition to this, the amount awarded is judged according to specific criteria such as the expected effect on traffic volume, profit or other services (see Avinor, 2008).

On 3 February 2004, the European Commission ruled that incentives granted to Ryanair by the Walloon government for the use of Brussels South Charleroi Airport, which were estimated to be worth €4.5 million, breached European Commission state aid rules. The ruling meant that state-owned airports in Europe could only offer incentives if they met the following criteria (Howrey Europe, 2004):

- They must be necessary to the opening of new routes.
- Their duration must be limited (maximum of five years for direct flights).
- They must not represent more than 50 per cent of the costs incurred to start the new route.
- They must be available to any airline established at the airport.
- They must be granted in a transparent and non-discriminatory manner.

The Commission ruling did not apply to privately owned airports. However, very few small regional airports are privately owned (Halpern and Pagliari, 2007). In addition, a ruling on 17 December 2008 from the Court of First Instance overturned the European Commission decision regarding incentives granted to Ryanair by the Walloon government for the use of Brussels South Charleroi Airport. As a consequence, the European Commission will need to review its criteria for airport incentives.

Place

Airports sell direct to airlines for rights to use the airport. Management processes are important here as communications between the airport and airline need to be efficient and effective. Some airports assign specific personnel to deal with airlines during the route development process. For instance, Aberdeen Airport, a large regional airport in Scotland, introduced an 'Entry into Service Process'. This consists of an assigned project manager who acts as a point of contact and provider of assistance to airlines establishing new routes or expanding existing services at the airport. This might not be necessary at small regional airports where the organisational structure may be less complicated than at larger airports. However, many small regional airports belong to a national or regional airport system and communication processes may be negatively affected by different levels of involvement in the development of air services. For instance, there might be a head office for the airport company that has responsibility for developing air services, the airport manager may be involved and there might also be a regional development group involved. Indeed, Halpern and Pagliari (2007) compared levels of market orientation (defined as the ability of an airport to understand, share and respond to the needs of airlines) and found levels to be significantly lower at airports operated as part of an airport system vis-à-vis airports operated as an independent entity.

Generally speaking, airports do not sell direct to passengers for rights to use the airport. Instead they rely on intermediaries such as airlines (including tour operators), travel agents or travel planning portals to reach end-users. However, airports are increasingly involved in providing online travel planning support to passengers (i.e. by providing online timetable services and links to airline websites, the tourism industry and travel planning portals). Figure 5.5 provides an example from Chambéry-Savoie Airport. The airport homepage shows the destinations that are served and provides links to airline websites, car hire and coach transfers.

More information is provided by clicking on the links at the top left corner of the homepage. For instance, the flights section provides timetable information, whilst exploring the region provides tourism information for the region. There is also a business aviation section that provides airlines with a range of information, including data on the catchment area and potential demand, information about the airport infrastructure and services, and details about the airport user charges and incentives scheme.

The provision of online travel planning support to passengers is important given the rapid growth in online sales in the airline industry and in the travel industry in general. For instance, online travel sales in Europe have increased from €0.2 billion in 1998 to €58.4 billion in 2008, with the market share of online travel sales increasing from 0.1 per cent to 22.5 per cent over the same period (Marcussen, 2009). Airlines provide the greatest contribution towards online travel sales in Europe, accounting for 55 per cent of all online sales in 2008 when sales by type of travel service is considered (this includes airlines, trains, car rentals, hotels and holiday packages). Airlines do still use other means of distribution but online distribution has grown rapidly in recent years, especially with low-cost airlines where the target is often to achieve 100 per cent online sales. As an example, online sales contributed 20 per cent of total sales for the low-cost airline Norwegian in 2002. This increased to 87 per cent by 2008 (Norwegian, 2003, 2009).

Figure 5.5 Homepage of Chambéry-Savoie Airport

Providing online travel planning to passengers is only one of a number of benefits that airports can achieve by having an online presence. They can also use the airport website to act as a wider tool for marketing to airlines – the business aviation section of the website for Chambéry-Savoie Airport provides a good example of this. There are also revenue-generating opportunities available to airports from selling their own services (i.e. car parking and shopping) online. The website can also act as a host agency by providing advertising space and links to other websites where commission might be made from subsequent sales on those websites – the website for Chambéry-Savoie Airport provides a good example of this, providing advertising space not only for its airlines but also for companies such as Intersport ski rental, National car rental and Trans'Neige coach transfers.

Conclusion

This chapter has investigated the role of marketing as a means of developing passenger air services at small regional airports in Europe. The chapter demonstrates the need for airports to understand the product/service needs of their target markets, supporting cost reduction where possible. Airports should use direct and aggressive means of communication regarding opportunities for the development of air services. Communications should be intelligence based and airports may benefit from collaborating with local tourism and regional development agencies in order to pool resources and provide an integrated approach to the development of air services. Airports should offer price incentives that demonstrate a willingness to share in the risk associated with establishing a new route or expanding an existing service. They might also want to offer incentives that encourage airlines to use the airport as a base airport for their aircraft. Airports should provide marketing support for joint advertising and promotional campaigns. These campaigns can be conducted in-house or via intermediaries such as airlines (including tour operators), travel agents or travel planning portals. Airports should also support the travel planning needs of end-users by providing online timetable services and links to airline websites, the tourism industry and travel planning portals. Developing an online presence may subsequently contribute to the wider marketing activities of airports and allow airports to generate additional sources of revenue.

References

ACI-Europe (2004), *The Social and Economic Impact of Airports in Europe* (Geneva: ACI-Europe).

ATAG (2008), *The Economic and Social Benefits of Air Transport 2008* (Geneva: ATAG).

Avinor (2008), *Start-Up Aid and Aid to Marketing*, <http://www.avinor.no/avinor/trafikk/40_Incentivordninger> accessed 20 July 2008.

Baidal, J.A.I. (2004), 'Regional Tourism Planning in Spain: Evolution and Perspectives', *Annals of Tourism Research*, 31:2, 313–33.

Brassington, F. and Pettit, S. (2007), *Essentials of Marketing*, 2nd edition (Harlow: Pearson Education).

Chambéry-Savoie Airport (2008), *Market Research*, <http://www.chambery-airport.com/rubrique.php3?id_rubrique=60> accessed 28 July 2008.

Doganis, R. (2006), *The Airline Business*, 2nd edition (London: Routledge).

Eden, P. (2005), 'Route Development Funds: what they are and how they work', in *4th Forum on Air Transport in Remoter Regions*, 24–26 May, Stockholm.

Eurocontrol (2008), Unpublished data extracted from STATFOR – Air Traffic Statistics and Forecasts (Brussels: Eurocontrol).

European Commission (1996), Decision Number 1692/96/EC of the European Parliament and of the Council of 23 July 1996 on Community Guidelines for the Development of the Trans-European Transport Network: Official Journal L228, 09/09/1996 P.0001-0104 (Brussels: European Commission).

Finavia (2009), Unpublished data extracted from Finavia's air traffic statistics (Vantaa: Finavia).

Finnish Civil Aviation Administration (2004), *Annual Report of the CAA* (Helsinki: Finnish Civil Aviation Administration).

Gloersen, E. (2005), 'The Importance of Airports in Securing Accessibility and the Role of Demographics', in *4th Forum on Air Transport in Remoter Regions*, 24–26 May, Stockholm.

Graham, A. (2003), *Managing Airports: An International Perspective*, 2nd edition (Oxford: Butterworth-Heinemann).

Halpern, N. and Niskala, J. (2008), 'Airport Marketing and Tourism in Remote Destinations: Exploiting the Potential in Europe's Northern Periphery', in A. Graham et al. (eds), *Aviation and Tourism: Implications for Leisure Travel* (Aldershot: Ashgate Publishing).

Halpern, N. and Pagliari, R. (2007), 'Governance Structures and the Market Orientation of Airports in Europe's Peripheral Areas', *Journal of Air Transport Management*, 13:6, 376–82.

Howrey Europe. (2004), 'Ryanair/Charleroi Decision – When are Incentives Granted to Airlines by Regional Airports Lawful?', *Howrey Client Alert*, February.

Kealey, S. (2004), 'Route Partnerships in Action', *Airline Business*, November.

Lauterborn, B. (1990), 'New Marketing Litany: Four Ps Passe: C-Words Take Over', *Advertising Age*, 61:41, 26–7.

Marcussen, C.H. (2009), *Trends in European Internet Distribution of Travel and Tourism Services*. (Bornholm: Centre for Regional and Tourism Research).

McCarthy, J. (1960), *Basic Marketing: A Managerial Approach* (Homewood: Irwin).

Norwegian (2003), *Annual Report 2002* (Fornebu: Norwegian).

Norwegian (2009), *Annual Report 2008* (Fornebu: Norwegian).

Pagliari, R. (2005), 'Developments in the Supply of Direct International Air Services from Airports in Scotland', *Journal of Air Transport Management*, 11:4, 249–57.

Papatheodorou, A. (2002), 'Civil Aviation Regimes and Leisure Tourism in Europe', *Journal of Air Transport Management*, 8:6, 381–8.

Reynolds-Feighan, A.J. (1995), 'European and American Approaches to Air Transport Liberalisation: some Implications for Small Communities', *Transportation Research Part A: Policy and Practice*, 29:6, 467–83.

Waterford Airport (2008), *Route Development*, <http://www.waterfordairport.ie/ns/content/view/38/85/> accessed 20 July 2008.

Williams, G. (2002), *Airline Competition: Deregulation's Mixed Legacy* (Aldershot: Ashgate Publishing).

PART B
Subvention Mechanisms

Chapter 6

European Experience of Public Service Obligations

George Williams

In order to enable governments to maintain essential air services, Articles 16, 17 and 18 of Regulation (EEC) No. 1008/2008 define a system of public service obligations (PSOs) which can be imposed on carriers operating on designated routes. In essence, the legislation allows Member States to impose a public service obligation in respect of scheduled air services between any airport in the Community and an airport serving a peripheral or development region within its territory or on a thin route to any airport in its territory considered vital for the economic and social development of the region served by the airport. If no airline is willing to provide a service under the conditions imposed, the government may restrict access to the route to a single carrier and award financial compensation to the carrier in return for compliance with the PSO. While a PSO can be imposed by a Member State for a route between its territory and that of another Member State, over 90 per cent of public service obligations are in respect of domestic services.

In the Czech Republic, Finland, Greece, Ireland, Portugal and Sweden, national government departments administer air transport public service obligations, while in France, Germany, Italy and Spain, administration is in the hands of regional authorities. In the UK, the Scottish Government is responsible for administering the routes operated from Glasgow and the respective regional authority for services provided in Orkney, Shetland and Western Isles, while in Wales it is the Welsh Assembly Government.

Use of the PSO

Eleven European Union (EU) Member States (Czech Republic, Finland, France, Germany, Greece, Ireland, Italy, Portugal, Spain, Sweden and the UK) currently provide subvention for a total of 142 public service obligations. In addition to these, Iceland, Norway and Switzerland[1] also impose public service obligations, making a total of over 200 in operation across Europe. A listing of all current EU public service obligations is provided on the Air Transport Portal of the European Commission.[2] Within the EU, France has the largest number of PSO

1 Switzerland issued an invitation for tenders to operate a PSO between Berne and Lugano in November 2007.

2 <http://ec.europa.eu/transport/air/internal_market/doc/pso_210109.pdf>.

routes, followed by Italy, Portugal, the UK and Greece, with between 25 and 30 each (Figure 6.1). In addition to these subsidised restricted access public service obligations, there are currently 76 open access public service obligations listed on the Commission's website. Not all of these however are being operated at present (Figure 6.2).

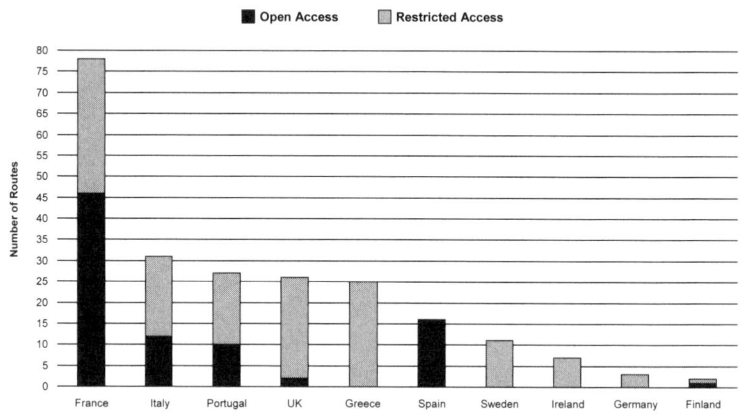

Figure 6.1 EU PSO routes by designating country (May 2007)
Source: DG TREN.

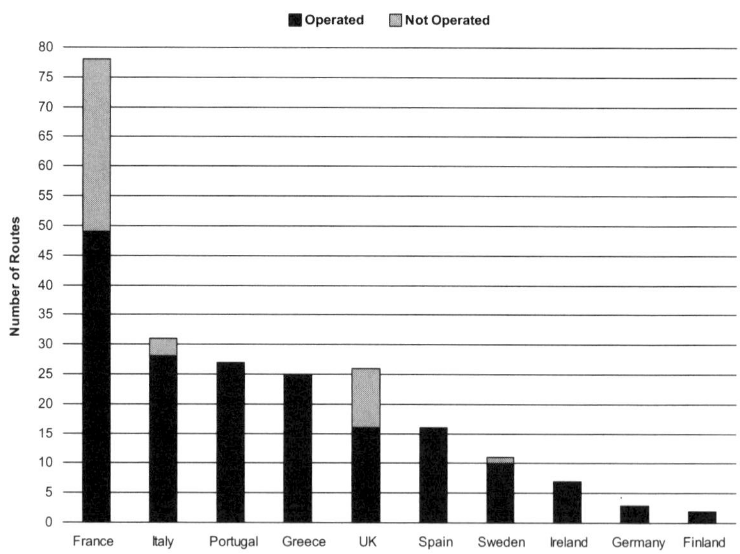

Figure 6.2 EU PSO routes operated and not operated (May 2007)
Source: DG TREN.

In terms of the types of routes that have been designated as public service obligations, there are 75 inter-island routes and 68 that link islands to the adjacent mainland. Of the remaining routes, 63 connect domestic mainland locations, 16 are international and four are domestic long-haul sectors linking France to its overseas territories (Figure 6.3).

Conditions Imposed

All restricted access public service obligations require the operating carrier to adhere to fixed levels of service for the duration of the contract. In the majority of cases, the airline is required to provide a minimum daily service frequency and/or number of seats. It is the responsibility of the administering authority to determine what this should be. There are often specific timetabling requirements with which the operator must comply. Many of these are there to enable passengers to undertake day return trips and to make convenient onward connections. In addition, there have been an increasing number of conditions set relating to the type of aircraft that must be employed. Table 6.1 contrasts the conditions set on air transport public service obligations in a selection of European countries.

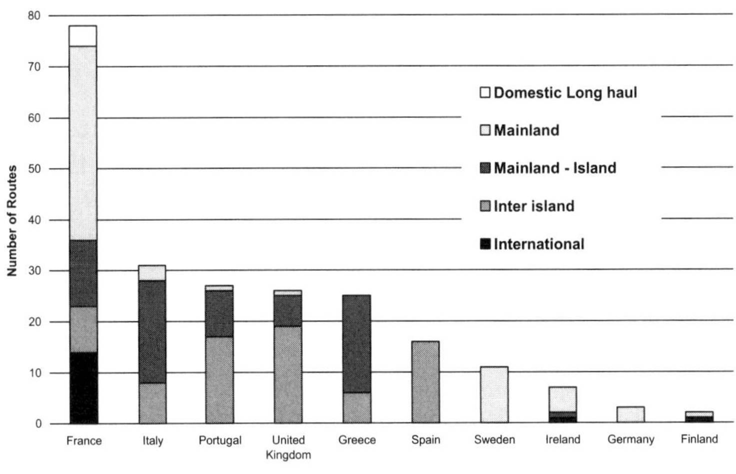

Figure 6.3 EU PSO routes by type (May 2007)

Source: DG TREN.

Table 6.1 PSO minimum operating requirements by country

Country	Minimum frequency (round trips)	Minimum capacity	Minimum aircraft size	Timetable requirements	Maximum one-way fare
France	2 per weekday	No	Yes	Yes	No
Germany	2 per weekday	No	Yes	Yes	Yes
Ireland	3 per weekday	Yes	Yes	Yes	Yes
Italy	4 per weekday	Yes	Yes	Yes	Yes
Norway	Up to 5 per weekday	Yes	Yes	Yes	Yes
Portugal	2 per weekday	Yes	No	Yes	Yes
Scotland	2 per weekday	No	Yes	No	Yes

Note: The requirements most commonly in use are shown.

France

As is standard practice across the EU, all French mainland PSOs stipulate a minimum level of service frequency, usually two round trips per day. A minimum aircraft size is required, ranging from a 9-seater on Brest–Ouessant to 60 seats on the majority of routes. On Paris–Périgueux, the operator must deploy pressurised aircraft. Most of the mainland PSOs in France also stipulate timetabling requirements. Precise timings though are not specified, but operators must allow for passengers to have at least a certain number of hours at their destination (usually a minimum of seven). On all PSO routes involving Paris Orly, the French Government has reserved slots at the airport for the services. Limits on the amount carriers can charge passengers are only stipulated in a small number of PSOs (e.g. Paris–Agen and Brest–Ouessant).

As regards PSOs between mainland France and Corsica, traffic on some of these routes is relatively large, particularly during the summer holiday season. All, however, require different minimum levels of service frequency and seat capacity that must be provided in the summer and winter seasons. The only minimum aircraft size requirements are on services from Montpellier to Ajaccio and Bastia, while on the Paris routes operators must use jet aircraft. With regard to regulation of air fares, there are limits on the amount carriers can charge for full economy one-way fares on all Corsican PSOs. In addition, there is a lower maximum fare for specific categories of passengers, such as students and senior citizens.

Germany

German PSOs stipulate a minimum level of frequency to be supplied by the operator, with the only capacity requirements in place on the Hof–Bayreuth–Frankfurt route. In all cases, minimum aircraft size and specific timetabling requirements are stipulated. For example, on the Erfurt–Munich route services must be scheduled between 06:00 and 21:00 in order that flights should arrive in Munich to enable connections to be made to onward destinations. Because of these requirements, the Federal Government has reserved departure and arrival slots in Berlin, Munich and Frankfurt.

Ireland

Three out of the five mainland PSOs in Ireland require that the air carrier operate at least three round trips per day. Due to low traffic, the requirement on the Dublin–Donegal route is one round trip per day. The PSOs also specify a minimum capacity to be supplied on each route and that aircraft should be pressurised and have a minimum seating capacity of 30. Air carriers are also required to meet various timetabling restrictions in three out of the five PSO contracts. While these do not specify precise departure and arrival times, some of the contracts demand an early morning departure from the regional airport and a late evening departure from Dublin to allow sufficient time for passengers to complete a day's business in the capital. At present, there are no departures and arrivals slots at Dublin reserved for PSO services. While carriers have the freedom to set their own fares, there are requirements that they offer a certain percentage of their capacity at specified levels.

Italy

On Italian PSOs, carriers must operate aircraft equipped with at least 140 seats. Timetabling requirements are also stipulated in the PSO tenders. The Italian Government reserves slots at Rome and Milan to secure departure and arrival times. Fare regulations are very similar to those imposed on the Corsican–mainland French routes, with a maximum one-way, fully flexible economy fare stipulated. Each PSO stipulates a minimum level of service frequency; on Cagliari–Rome, for example, the requirement is for eight round trips per day. Air carriers must also supply a minimum level of frequency, usually expressed in terms of a daily supply of seats.

Norway

PSOs in Norway have been imposed on over 60 routes, divided into 18 separate tender groups. Due to remoteness and the low traffic levels associated with many of Norway's regional airports, the government decided to arrange tenders in groups to enable tendering carriers to achieve economies of scale and scope by operating networks rather than individual sectors. The terms of most of the PSOs are fairly complex, involving a range of frequency, capacity, aircraft size and timetabling requirements. Most specify a certain number of round trips per weekday, with normally a separate requirement for weekends. The vast majority of tenders stipulate a pressurised aircraft equipped with a minimum of 30 seats.

In addition, timetabling requirements are fairly stringent, with flights having to be scheduled at specific times. The regulation of fares is fairly straightforward, with maximum one-way fares being stipulated.

Portugal

Aside from one route linking Lisbon with Villa Real and Braganca, most Portuguese PSOs link the mainland (Lisbon and Porto) with airports in the Azores archipelago (Ponta Delgada, Horta, Pico, Santa Maria, Terceira) located some 1400–1800 km west of the mainland. PSOs are also imposed between Funchal (Madeira) and Ponta Delgada, Funchal and Porto Santo, and Lisbon and Porto Santo. Inter-island services connecting all nine Azorean islands are also subject to a PSO.

There are differences in the minimum service requirements contained in Portuguese PSOs. While all stipulate a minimum level of service frequency, some require the operator to maintain a schedule based on a minimum level of capacity. A number of the PSOs do not set a minimum aircraft size or specify a type of aircraft; however services connecting the mainland with the Azores require the operator to use jet aircraft. Timetabling requirements are also in place in a number of cases. For example, on Portuguese mainland–Azores PSOs, flights must take place between 06:30 at the place of departure and 12:30, while at least one flight per day between Lisbon and Ponta Delgada and between Lisbon and Terceira must be operated between 08:00 and 21:00; and for at least three days a week, one of these must be operated before 13:00.

Scotland

All Scottish PSOs stipulate a minimum level of service frequency, which bears some relation to the annual traffic on each route. For example the busiest route, Glasgow–Campbeltown, has a minimum frequency requirement of two round trips per weekday, while less busy routes in the Shetland Islands require in some cases two round trips on a Tuesday or a Saturday. The Scottish Government PSOs require a minimum capacity of between 11 and 15 seats per route per day, while in all other cases the only capacity requirement is that the routes should be operated by a twin-engined aircraft with a minimum capacity of eight seats. None of the PSOs specify any timetabling requirements. However, all stipulate a maximum one-way single fare.

Fares Charged on PSO Routes

Most public service obligations stipulate the maximum fares that may be charged. Wide variations exist though in these levels for routes of similar distance and traffic volumes, with certain countries being much more generous to travellers in terms of setting lower maximum fares and paying higher levels of subvention (Williams and Pagliari, 2004). Figure 6.4 compares the highest return fares charged in 2007 on a representative selection of PSO routes operated in France, Greece, Ireland, Italy, Portugal, Spain, Sweden and the UK.

PSO Tendering Process

It is clear that established carriers, especially those with previous experience of operating routes in peripheral regions, have an undoubted advantage over new entrants. Figure 6.5 reveals that most public service obligations are operated by locally based carriers. Even though tenders until very recently covered only three years[3] of operation, several carriers pulled out of their contracts early. Most did so as a result of underestimating the costs of operating the service they had been contracted to provide. In a few instances the tender-winning carrier entered into bankruptcy during the course of the contract.[4]

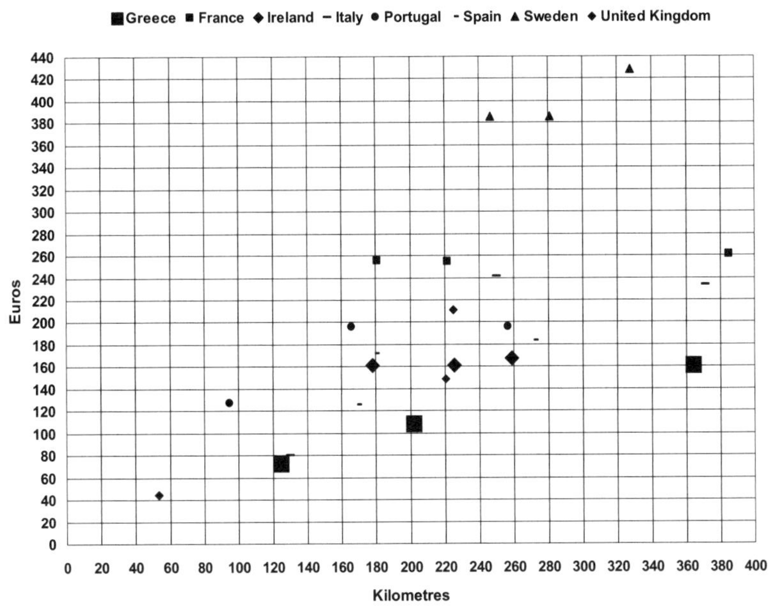

Figure 6.4 Comparison of highest return fares on selected PSO routes (May 2007)

3 In its revisions to the Third Package the Commission has increased this to four years and in the case of ultra-peripheral regions, such as the Azores, to five years (Regulation EC No. 1008/2008).

4 In 2002 this fate befell Euroceltic, which had been operating two PSO routes in Ireland (Dublin–Donegal and Dublin–Sligo).

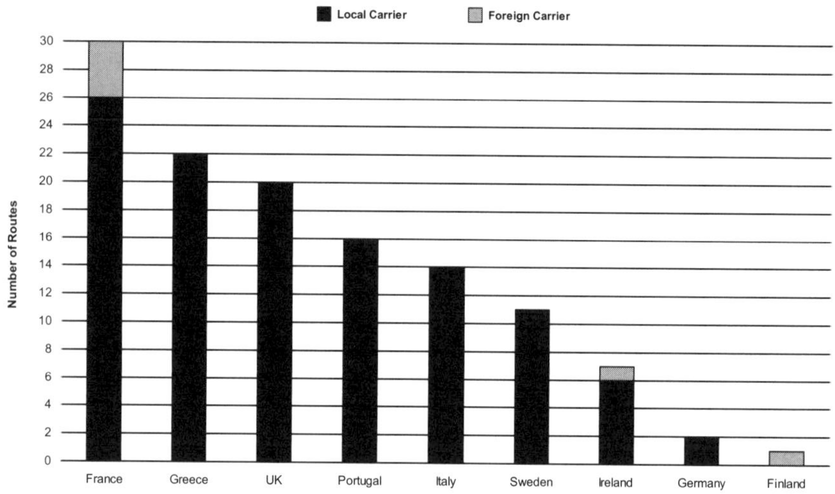

Figure 6.5 EU PSO routes by origin of operator (May 2007)
Source: DG TREN.

Partly as a result of these experiences with the tendering process, the operation of PSO routes rests mainly in the hands of tried and tested carriers. By way of example, Tables 6.2 and 6.3 show the results of the 2002–2005 and 2005–2008 tendering rounds in Ireland. The comparatively short time allowed between notification that a carrier's tender has been selected and when it must commence operation has been cited by some prospective airlines as a deterrent to submitting a bid (O'Fee, 2003). This is particularly the case when a carrier based in one Member State is contemplating a PSO bid on a route in another country. The leasing of aircraft, the setting up of an operating base and the arranging of crewing, ground handling, maintenance etc. all take time and may represent a significant outlay to a small carrier. Local bureaucratic procedures have also been cited by prospective carriers as an impediment to their tendering for a route. In the event of not winning a tender round, airlines may be faced with significant sunk costs. It is perhaps not surprising therefore that most PSO routes continue to be operated by locally based airlines.

Figure 6.6 shows the number of restricted access PSO routes operated by each carrier. In all there are 28 airlines operating services, with Olympic Airlines leading the field with 22 routes.

Table 6.2 Irish PSO tendering round 2002–2005

	Previous operator	Successful tender	Current operator
Dublin–Derry	Loganair	Loganair	Loganair
Dublin–Donegal	Aer Arann	Euroceltic*	Aer Arann
Dublin–Galway	Aer Arann	Aer Arann	Aer Arann
Dublin–Kerry	Aer Arann	Aer Arann	Aer Arann
Dublin–Knock	Aer Arann	Aer Arann	Aer Arann
Dublin–Sligo	Aer Arann	Euroceltic*	Aer Arann

Source: Department of Transport (2003). *Euroceltic ceased trading in winter 2002.

Table 6.3 Irish PSO tendering round 2005–2008

	Previous operator	Successful tender	Current operator
Dublin–Derry	Loganair	Loganair	Loganair
Dublin–Donegal	Aer Arann	Aer Arann	Aer Arann
Dublin–Galway	Aer Arann	Aer Arann	Aer Arann
Dublin–Kerry	Aer Arann	Aer Arann	Aer Arann
Dublin–Knock	Aer Arann	Aer Arann	Aer Arann
Dublin–Sligo	Aer Arann	Aer Arann	Aer Arann

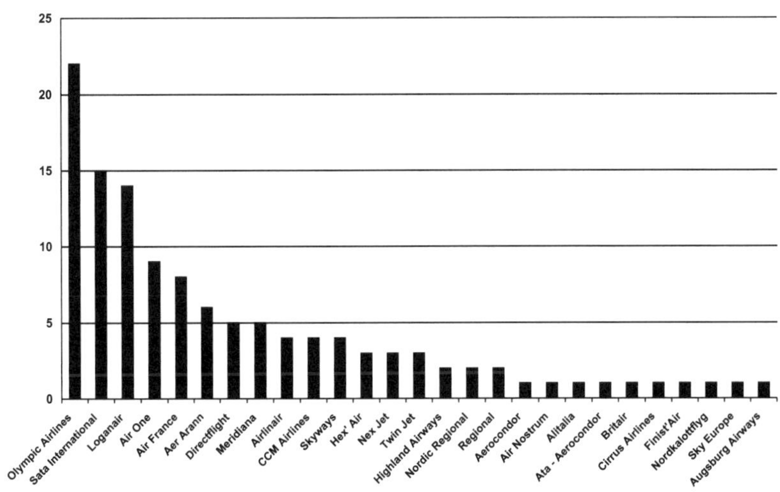

Figure 6.6 Restricted access EU PSO routes by designated operator (May 2007)

Types of Aircraft Used on EU PSO Routes

Figure 6.7 gives a breakdown by country of the seating capacities of aircraft used on PSO routes. It is clear that there are large differences across Europe, with a large proportion of such routes in France, Italy, Portugal and Spain operated by jet aircraft seating over 60 passengers, whilst in the UK most PSO services use turboprop or piston equipment with fewer than 20 seats. This in part reflects the different volumes of traffic carried and the distances flown. The average PSO route distance by country is shown in Figure 6.8. Aside from the routes connecting France with its overseas territories, the longest PSO sectors link mainland Portugal with the Azores. Services to the Azores are operated by SATA and TAP mainly using Airbus 310 and Airbus 320 aircraft. The large hold capacity of the wide-bodied A310 is of particular importance for these services. The distances flown are over 1,400 km, with a flight time of around two hours. Airbus 320 aircraft seating over 150 passengers are also normally used between Paris (Orly) and Ajaccio (Corsica) by Air France, but during summer traffic peaks the carrier has regularly used Boeing 747-400 equipment with over 400 seats available (Minshall, 2004).

At the other end of the scale the inter-island service linking Westray and Papa Westray in Orkney involves a distance of only 2 km and a flight time of just two minutes. The route is flown by Loganair using BN Islander aircraft equipped with nine seats. Figure 6.9 shows this type of aircraft in operation at Papa Westray.

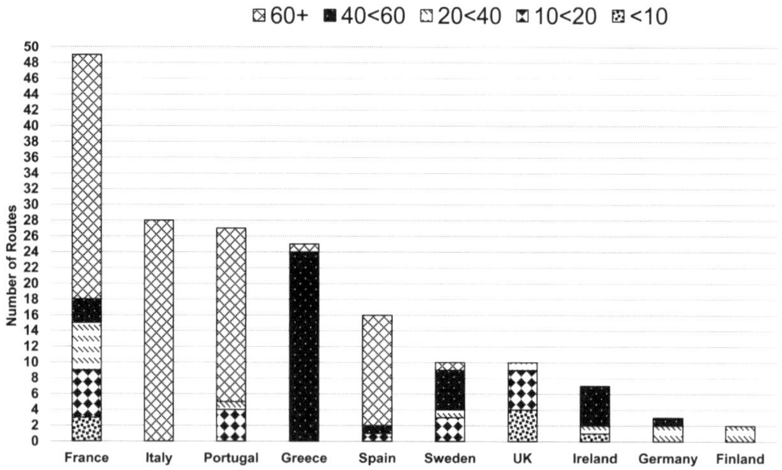

Figure 6.7 Seating capacity of aircraft used on EU PSO routes (May 2007)

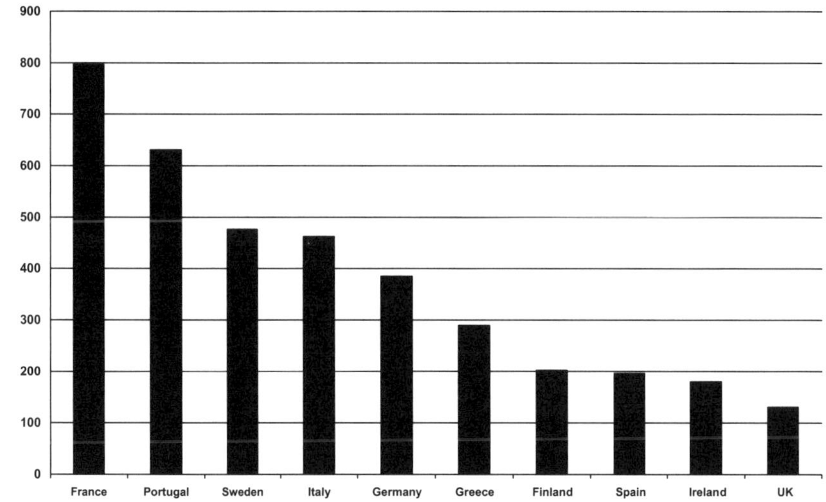

Figure 6.8 EU PSO routes by average distance in km (May 2007)

Figure 6.9 BN Islander of Loganair at Papa Westray, Orkney Islands

Levels of Subvention

Subvention levels on individual PSO services have been on the rise, in some instances dramatically so. Ireland, for example, has seen annual subsidy levels on four of the country's PSO routes, Dublin–Donegal, Dublin–Galway, Dublin–Kerry and Dublin–Sligo (see Tables 6.2 and 6.3), rise from 4.2 million euros in

2000 to 14.4 million euros in 2003[5] (Department of Transport, 2003). While the level of service provided has improved significantly,[6] it is not realistic to attribute the hugely increased amount of subvention solely to this. The amount of subsidy required per one-way journey varies widely between routes, in the case of these four Irish PSO routes in 2003 from 53 euros on Dublin–Galway to 290 euros on Dublin–Knock. In the current economic difficulties facing Ireland the country's Finance Ministry has released a report that raises the possibility of terminating the funding of its PSO routes.[7]

Some more recent data from Sweden also reflects the wide variation between PSO routes in the amount of subsidy provided for a one-way journey. In 2006, the subsidy per one-way trip between Stockholm (Arlanda) and Gällivare amounted to 30 euros, whilst between Pajala and Lulea the equivalent amount was 280 euros, reflecting the much lower volume of traffic on this route (see Chapter 7).

In 2003, the operators of the four PSO routes linking Paris (Orly) with Corsica (Ajaccio, Bastia, Calvi and Figari) received a subsidy of 18.3 million euros to carry close to 1 million passengers.[8]

Variation in the Interpretation of the Justification for PSO Intervention

At a fundamental level, a number of Member States and other interested parties have claimed that the criteria for intervention, i.e. for designating a route a PSO, are unclear. The UK Government, in particular, has voiced concern at the interpretation of the terms used to define the conditions under which a PSO could be imposed (Department for Transport, 2004). While many of the states implementing PSO legislation have adopted a very conservative approach, nominating only the most vulnerable of air services for the type of protection offered under the legislation, others have taken a much less restrictive stance. The more liberal attitudes of some governments to the interpretation of PSO conditions are manifest in the number of routes so nominated and in their particular characteristics.

Subsidy is supposed to be not eligible 'where other forms of transport can ensure an adequate and uninterrupted service'. It is clear however that there are a number of PSO routes currently in operation that flout this condition. For example, the Dublin–Galway route, which generates around 100,000 passengers annually, has good surface public transport available. Aside from the four rail services a

5 The annual level of subvention for all six Irish PSO routes in the 2005–2008 tender round was set at 15.3 million euros.

6 In particular, larger capacity pressurised aircraft are now used.

7 Report of the Special Group on Public Service Numbers and Expenditure Programmes, vol. 1, p. 74, July 2009, Ireland.

8 Zuccarini-Gagnoni (2002) estimated that the total amount of subvention for all French PSO services in 2001 amounted to €200 million.

day, with a journey time of three hours, there is a frequent express coach service (Williams, 2005).

The disparity of approach is clearly seen in the extremes of capacity provision on routes operated under the protection of PSO. While some of these have daily flights by small turboprop aircraft, others feature multiple flights each day with large jets. For example, the number of passengers carried on the PSO route between Paris (Orly) and Ajaccio in 2006 was over 435,000, whilst just 225 passengers travelled in the same year between Lerwick and Papa Stour in the Shetland Islands.

Critics of the current PSO system have suggested that some of the largest of these routes could be self-sustaining, without recourse to subsidy or the protection of monopoly status. Social assistance in such cases might better be handled by offering residents of peripheral regions assistance in the form of direct discounts on ticket prices, rather than through subsidising airline operations. A scheme of this nature is in force for transport between the Madeira archipelago and mainland Portugal, and the Scottish Executive received approval in 2006 for the operation of a similar scheme in the Scottish Highlands and Islands.[9]

Low-cost Provision of Services on Essential Routes

Low-cost carriers seeking opportunities in the European market have focussed attention on a number of apparent inconsistencies resulting from the different approaches to PSO intervention. Clearly, where a PSO is in force the relevant national authority is able to refuse a second carrier access to the route or routes involved. This has effectively prevented low-cost carriers easyJet and Ryanair from entering markets in France and Italy where they felt they could offer services without the need for subvention. For example, in 2005 easyJet was forced to cancel plans to operate Olbia–Milan and Ryanair had to pull off Alghero–Rome, a route it had been operating for a year.

The Commission subsequently opened a formal investigation into the rules imposed by Italy in May 2006, which involved the bundling of 16 domestic PSO routes linking three airports in Sardinia with cities in mainland Italy and Sicily. Prior to this there were six PSO routes in operation between Sardinia and the mainland/other parts of Italy that were tendered for separately. This new PSO regime required potential operators to enter the market at a much greater scale than they would perhaps have desired, effectively creating a significant barrier to entry. In April 2007, however, the Commission decided to authorise Italy's introduction of PSOs on the 16 routes linking Sardinia with mainland airports, subject to certain

9 Scotland's Air Discount Scheme provides residents of Orkney, Shetland, the Outer Hebrides, Islay, Jura, Caithness and Sutherland with a 40 per cent reduction on their air fares on non-PSO routes within Scotland. <http://www.airdiscountscheme.com/airds/CCC_FirstPage.jsp>.

conditions in order to avoid restrictions on the launch of new services. The impact of withdrawing PSO status on routes linking Sardinia with the Italian mainland were explored in great detail by di Francesco, who concluded that there was a strong case for their retention (di Francesco, 2008).

Administering PSOs

In view of the fact that PSO subsidies have increased over time, some substantially so, providing incentives to operators to reduce their costs and to innovate should become a feature of the tendering process. Those responsible for administering PSOs need to use appropriate benchmarks by which to judge whether bids are soundly based or not. The apportionment of an airline's operating costs to particular services is to a certain degree a matter of subjective judgement, particularly with regard to the indirect elements in the cost mix.[10] A carrier making what it believes to be the only bid for a PSO route may well be tempted to allocate a disproportionate amount of the company's costs to the subsidised operation in the knowledge that it is unlikely to be subject to any serious degree of scrutiny.

Greater transparency is needed in respect of the levels of subsidy paid to maintain PSO services. Several EU Member States are unwilling to divulge how much they are paying carriers to maintain their PSO operations. There can be no justification for this policy, which simply encourages the notion that there is something to hide. Either the PSO system is fully open and transparent, or it is not. On the basis of the available evidence, at present it is not possible to conclude the former.

References

Department for Transport (2004), *Consultation on the Protection of Regional Air Services to London*. Stationery Office, London.

Department of Transport (2003), *Review of Air Services Supported by the Essential Air Services Programme*. Report by DKM Consultants (Dublin).

Di Francesco, A. (2008), *The Potential Impact of Removing Public Service Obligation Regulations on the Air Transport Market between Sardinia and the Italian Mainland*. Unpublished MSc thesis, Cranfield University.

Minshall, C. (2004), *Use and Reform of the Public Service Obligation*. Unpublished MSc thesis, Cranfield University.

O'Fee, B. (2003), *Tendering for a PSO Route: The Experience of a New Entrant Airline*. Third Forum on Air Transport in Remoter Regions (Cork).

10 An airline's indirect operating costs comprise items of the cost mix not directly related to the operation of aircraft.

Williams, G. (2005), European Experience with Direct Subsidization of Air Services, *Public Money & Management*, 25, 3, 155–61.

Williams, G. and Pagliari, R. (2004), A Comparative Analysis of the Application and Use of Public Service Obligations in Air Transport within the EU. *Transport Policy*, 11, 55–66.

Zuccarini-Gagnoni, L. (2002), *A Comparison of European and US Approaches to Subsidising Air Carriers*. Unpublished MSc thesis, Cranfield University.

Chapter 7
Experience of PSO and the Tendering Process in Sweden

Göran Anger, Johan Holmér and Pär-Erik Westin

Introduction

Airline deregulation created a need for public service obligation (PSO) legislation in the European Union (EU) air transport reform packages. Thin regional routes, which previously survived thanks to cross-subsidisation from large monopoly routes, had little or no chance of survival under commercial conditions. In Sweden, cross-subsidisation did continue after deregulation in order to maintain a hub-and-spoke network structure, however generally with much reduced service quality to the remoter regions of Sweden. Arguably, air transport was and still is a necessary contributor to the social and economic development of many rural regions in Sweden. When accessibility was challenged, politicians were asked to impose government support in order to promote sustainable air transport services from the periphery to the centre of the country (Stockholm and beyond). This chapter examines the process and scope of PSO intervention in Swedish air transport. It also analyses the experiences of three major tendering processes and their disparate outcomes – and their lessons to learn for the future.

The Swedish Domestic Air Transport Market Then and Now

The Infrastructure

It was not until the 1960s that Sweden had a domestic air transport network. This was partially due to the fact that Sweden only had ten functional airports for regular air transport by 1960, but also because incumbent SAS was focusing on international flights and a few domestic trunk routes. During the 1960s and 1970s Sweden witnessed a heavy airport construction process thanks to political awareness of its regional wealth impact, not least in the remote regions. By the mid-1980s Sweden had more than 40 airports serving the domestic network. The construction of the most recent regional airport, Pajala in Swedish Lapland, was completed in 1999.

The System of Cross-Subsidisation

Before deregulation, Swedish domestic route distribution was based on a system of precedence, where SAS had the concession on domestic routes with remunerative prospects. This policy created three parallel domestic network systems, with SAS holding the primary network (major routes); its cross-subsidised subsidiary Linjeflyg the secondary network (mid-size routes) and finally a tertiary network served by regional commuter airlines. The tertiary route network was dependent on cross-subsidisation from either SAS or Linjeflyg and was normally operating under contracted feeder arrangements for SAS/Linjeflyg. Thus, the government authorised market precedence and the monopoly rights were given under condition that the carriers allowed for wealth distribution through cross-subsidisation. As a result, the regional air transport network prospered and Sweden saw a sharp increase in domestic air transport, especially in the 1980s when the number of passengers doubled. Two driving growth factors were the establishment of a single hub-and-spoke system in Sweden and a more effective yield management approach, which for the first time permitted discounted fares for leisure travel. To conclude, from the mid-1960s until 1990, Swedish domestic air transport witnessed an annual increase of 10 per cent in terms of passenger numbers. Economic growth was the overall driver behind this impressive rise in traffic, thus this development was similar to that experienced in the rest of Europe.

Airline Deregulation and Its Effects on Domestic Regional Air Transport Distribution

With the multilateral liberalisation of air transport (the Third Package) in the EU in 1993, Sweden immediately joined the agreement as a European (EES) member. The subsequent free market entry, capacity and fare setting altered market conditions in the Swedish domestic air transport, in theory at least. In reality little changed for a decade in terms of market share. SAS continued to dominate all trunk routes and instead of its previous partner, Linjeflyg, which was acquired by the SAS Consortium in 1991, regional commuter Skyways continued to feed SAS Swedish hub Stockholm Arlanda with minor regional spokes; in fact the number of domestic routes has increased sharply since deregulation, thanks to regional commuters. This time, however, cross-subsidisation was not eligible since each regional route in the domestic network had to stand on its own commercial merits. No state support for essential air services was considered by the government on the eve of deregulation.

Factors Behind the Domestic Market Stagnation in Sweden

Since 1990 the number of Swedish domestic air passengers has decreased by a quarter and more than half of all domestic routes have declined in passenger volume. Over 40 airports are still served by scheduled air transport; however the total number of annual domestic passengers is only 7 million, which places Sweden in 7th position in Europe, far behind countries such as Norway.[1] Deregulation cannot be blamed for this development; again the primary reason can be found in a long-term economic recession in Sweden and a number of socio-economic changes. Heavy rail and road infrastructure investments have also played a significant part in the change of travel demand in Sweden, which in turn have led to the closure of a number of airports in southern Sweden. However, although major routes have experienced competition and decreased fares, the majority of domestic routes, especially minor regional routes, have seen a sharp increase of fares in order to maintain profitability (Figure 7.1).

The Swedish Civil Aviation Authority (CAA) predicts a continued decline in the domestic market; until 2020 the aggregated forecast is a passenger volume drop of 22 per cent from today's numbers. The authority blames long-term structural changes, such as increased competition from high-speed trains and new motorway investments. Whereas domestic air travel in Sweden has gradually declined, passenger volumes on the PSO routes are generally more buoyant.

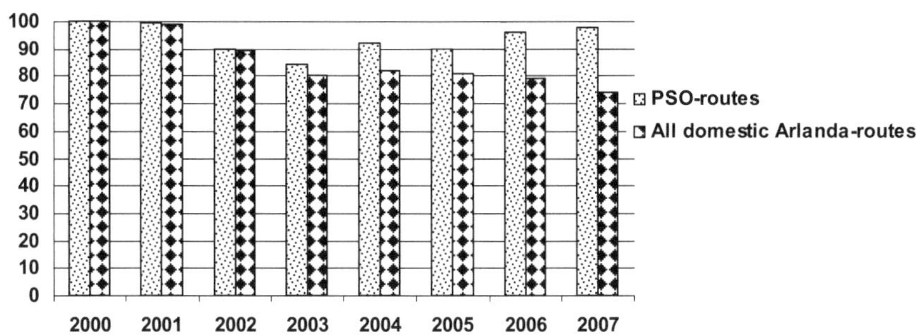

Figure 7.1 Index of passenger traffic volumes

Sources: Rikstrafiken and Luftfartsstyrelsen.

1 The number of annual flights per capita in Norway is 4 and in Sweden 2.5 according to Eurostat. In absolute numbers, Swedish domestic capacity (OAG, available seats) is about 1/3 less than in Norway and about half of UK capacity.

From Regulation to Deregulation, Air Transport Stands on Its Own

Although the Swedish government acknowledged the model of cross-subsidisation from major routes to finance minor routes in the domestic air transport network, the Swedish airline industry has generally never enjoyed public funding. While the airport industry was developed by various measures of public expenditure and airport cross-subsidisation, roads and railways were financed solely by tax funds. As deregulation commenced, the airline industry functioned satisfactorily under competition and accessibility to small communities was viewed to be well served without public intervention. This was generally governmental policy for a decade of deregulation experience, a decade when the domestic market decreased by a quarter followed by stagnation.

Processing PSOs and the Tendering Thereof: 2002–2008

The Establishment of a Specialised Government Agency

In a vast and sparsely populated country like Sweden one could expect that a PSO mechanism – as introduced by Regulation (EEC) 2408/92 – would find immediate practical use. Indeed, the premier set of PSOs (applying only to the single route Ostersund–Umea) was implemented already in 1994. As mentioned previously, the Swedish government kept a low profile in its effort to sustain frequent and coherent air services to remote regions during the entire first decade of liberalisation. The justification for imposing PSO on the Ostersund–Umea route had little to do with the prospects of socio-economic development as described in Regulation (EEC) 2408/92. Furthermore, the PSO itself was limited to simply two conditions, namely that two roundtrips per weekday (except Saturday and Sunday) be carried out and that the aircraft be equipped with lavatory facilities. Except for this single intervention on the part of the national authorities, it was left to the regions themselves to take any measures at their own discretion to sustain adequate air services. As those services were increasingly disintegrating or disappearing entirely, local authorities took action. In the absence of any national government initiative they embarked on a procurement exercise, in much the same way as they had always catered for various services in the community.

In retrospect the organisational setup in Sweden during the first phase of liberalisation shows several signs of immaturity and lack of clear direction. There was no national authority designated to analyse the need for PSO or to pursue tender procedures, let alone to compensate air carriers for the burden of PSO under contractual terms. Nor was there any national strategy elaborated on how or when to make use of the new mechanism at hand. Hence the provision of regular and adequate air services between the national hub Stockholm/Arlanda and remote regional airports was, in all practical aspects, a responsibility for local authorities and municipalities. As a result, municipalities in northern Sweden procured regular air services in order to achieve sustainable accessibility. However, no PSO had been imposed prior to procurement. Thus, Swedish national procedures failed to comply with Regulation (EEC) 2408/92. As a result, there is no official record of what amount of public spending had been necessary in Sweden to sustain adequate air services during this phase. Finally, on the one and single route where a PSO actually did apply, there was no supervision of compliance at all.

In 1998, the Swedish Parliament decided that a new government agency – Rikstrafiken – be established, the purpose of which was to engage in all modes of public transportation to be responsible for contracts necessary to sustain adequate long-distance transportation. In this process nothing was mentioned about the illegitimacy of the prevalent situation as regards air services to remote regions. Rikstrafiken's instructions made it quite clear that the agency had the right to engage and intervene in the markets for maritime transport, long-distance rail and coach services to maintain adequate transportation where need be. It was not as clear that the agency had the legal right to intervene likewise in the market for air services. Rikstrafiken made its first move in 2001 when it proposed that PSO be imposed from October 2002 on the ten routes (see Figure 7.2) where local authorities had already procured services for a decade, and that any compensation requested by air carriers from then on be financed by the agency (i.e. by state appropriation). This proposal was generally appreciated by the affected municipalities and endorsed by other actors in the market, e.g. the regulatory authority for civil aviation. For the first time since market liberalisation in 1992, Sweden had a body with a suitable mission and a will to apply the mechanisms of Regulation (EEC) 2408/92 in a rightful manner.

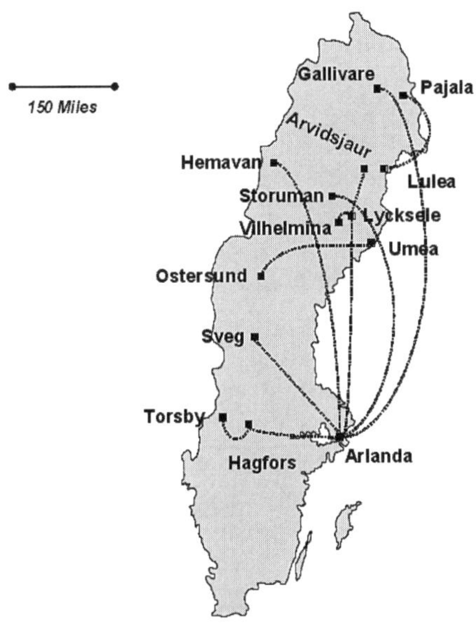

Figure 7.2 Swedish PSO routes 2002–2008

The establishment of the national agency Rikstrafiken was nothing less than a paradigm shift, inter alia because the nature of the contracts offered by Rikstrafiken would differ substantially from those previously offered by the municipalities. Rikstrafiken was to offer service concessions taking into account costs as well as revenue generated by the service, thus transferring business risks largely to the air carriers whilst at the same time giving air carriers commercial incentives to develop the services.

Gaining Hands-on Experience of PSOs and Tendering: Rikstrafiken's First Contract Round

Having analysed accessibility and alternative transport facilities, Rikstrafiken submitted a formal report to the Cabinet in October 2001 on the justification for PSO on ten different routes, nine of which were bound for the national hub Arlanda. The main argument was that only regular air services could provide these ten remote regions with transport facilities that would enable people to make a roundtrip to or from Stockholm on one and the same day. Failing this accessibility, Rikstrafiken argued that local industry would suffer and thus socio-economic development in the regions would be hampered. Having said this, it is clear that Rikstrafiken had not yet made any in-depth analysis as to the validity of these arguments and there was no model developed for the purpose. Rather, it was a case of simply performing the procedure that was expected by the Commission in a case like this. A further argument for imposing PSO was that the situation with local procurement, without PSO imposition, did not conform to Regulation (EEC) 2408/92.

In drafting the PSO, Rikstrafiken aimed at maintaining the quantity and quality of air services catered for by the local authorities, whilst at the same time harmonising conditions with those prevailing in other EEC countries at the time. Table 7.1 is a summary of the essential PSO requirements as originally proposed by Rikstrafiken.

In December 2001, the government decided that these requirements be imposed and in January 2002 commissioned Rikstrafiken to administer the related tendering procedure. At this early stage of development any contract that Rikstrafiken proposed to sign was subject to Cabinet approval, except where the Cabinet itself had explicitly commissioned Rikstrafiken to close a contract on a specific route. This meant that Rikstrafiken did not yet act as an independent and sovereign body.

Following discussions with the Commission on procedural aspects, Rikstrafiken submitted a formal written justification for PSO and made a visit to the Commission to further pave the way for the proposed actions. Having outlined the justification, including expected volumes of air services, geographical conditions and the lack of alternative modes of transportation, there appeared to be little doubt that the threshold for intervention applied by Rikstrafiken was going to be high by European standards. In other words, a region must indeed be very dependent on

Table 7.1 PSO requirements 2002–2005

Minimum seat capacity per year	Minimum frequency per week	Aircraft requirements	Timetable requirements	Maximum one-way fare*
Arlanda–Hagfors (ARN–HFS) 246 km				
7,175	2 per day (Mon–Fri)	Pressurised 15 seats	Yes	€189
Arlanda–Torsby (ARN–TYF) 280 km				
5,075	2 per day (Mon–Fri)	Pressurised 15 seats	Yes	€189
Arlanda–Sveg (ARN–EVG) 325 km				
11,550	2 per day (Mon–Fri)	Pressurised 15 seats	Yes	€213
Arlanda–Hemavan (ARN–HMV) 687 km				
10,700	1 per day Tue, Thu, Sun	Pressurised 48 seats	Yes	€232
Arlanda–Storuman (ARN–SQO) 591 km				
21,150	2 per day (Mon–Fri) 1 per day (Sat–Sun)	Pressurised 48 seats min	Yes	€232
Arlanda–Gällivare (ARN–GEV) 843 km				
76,700	2 per day (Mon–Fri) 1 per day (Sat–Sun)	Pressurised 48 seats min	Yes	€271
Arlanda–Lycksele (ARN–LYC) 546 km				
39,300	2 per day (Mon–Fri) 1 per day (Sat–Sun)	Pressurised 48 seats min	Yes	€232
Arlanda–Arvidsjaur (ARN–AJR) 663 km				
43,500	2 per day (Mon–Fri) 1 per day (Sat–Sun)	Pressurised 48 seats min	Yes	€245
Arlanda–Vilhelmina (ARN–VHM) 649 km				
25,500	2 per day (Mon–Fri) 1 per day (Sat–Sun)	Pressurised 48 seats min	Yes	€232
Luleå–Pajala (LLA–PJA) 192 km				
5,250	2 per day (Mon–Fri) 1 per day (Sun)	Pressurised 15 seats min	No	€74

*1€ = SEK 9.4

regular air routes before measures according to Regulation (EEC) 2408/92 were to be taken.

PSO routes account for approx. 3 per cent of Swedish national domestic air travel (see Figures 7.3 and 7.4). The 11 airports served from 2002 account for about 25 per cent of all airports in Sweden, be they national or regional.

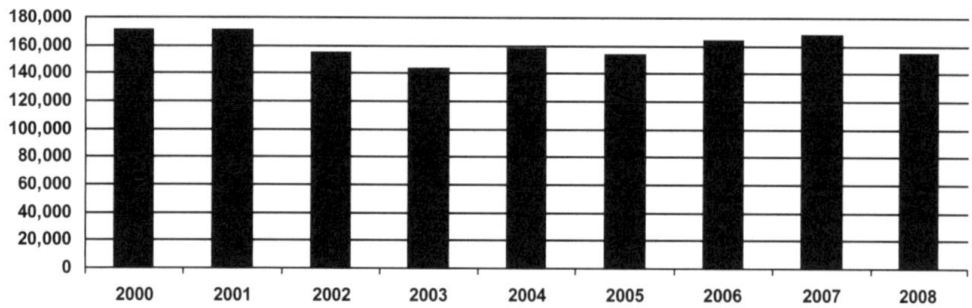

Figure 7.3 Development of total passenger volume per year on Swedish PSO routes

Sources: Rikstrafiken and Luftfartsstyrelsen.

Presumably, the aftermath of 9/11 in 2001 and Rikstrafiken's poor performance in the first contract round contributed to falling passenger volumes in 2002 and 2003.

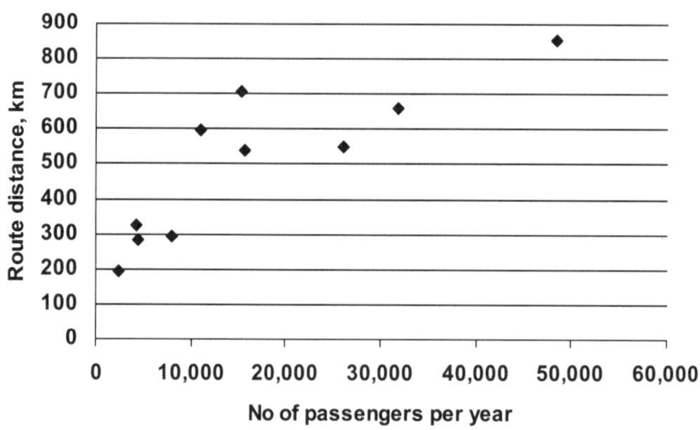

Figure 7.4 Relationship between route distance and number of journeys per annum (2007)

It was not until May 2002 that Rikstrafiken's first tender invitation was published in the EU Official Journal (OJ). This was late considering the inexperience and the fact that air services were to commence by the end of October that same year. Air carriers had but 31 days to submit a tender according to the details outlined in the tender dossier. Hardly any efforts had been undertaken by Rikstrafiken as regards market information prior to the tender. In June 2002, the single largest air carrier in Sweden – Skyways Express – publicly denounced Rikstrafiken's tender procedure because of the conditions defined by the agency, the main reason being the short period between the tender expiry date and the start of operations. This was a significant setback and bad publicity for Rikstrafiken during its first tender procedure. Despite these poor pre-conditions, five air carriers did submit tenders. Table 7.2 summarises the tender result.

Table 7.2 Tender result 2002

Route	Bidding air carrier/aircraft	Requested compensation (000€)/yr*
Arlanda–Hagfors/Torsby	Danish Air Transport/Beech 1900	1,200
	Wermlandsflyg/Jetstream 31-32	450
	AB Värmlandsflyg/Beech 1900	500
	European Exec. Express/Jetstream 31-32	950
Arlanda–Hemavan	Danish Air Transport/ATR 42	450
	AB Värmlandsflyg/ATR 42	1,300
	Norwegian Air Shuttle/Fokker 50	700
	European Exec. Express/ATR 42	1,550
Arlanda–Gällivare	Danish Air Transport/ATR 42-72	1,700
	AB Värmlandsflyg/ATR 42	2,050
	Norwegian Air Shuttle/Fokker 50	1,150
	Nordic East Airlink/MD 82-81	2,850
	European Exec. Express/ATR 42	3,400
Arlanda–Storuman/Vilhelmina/Lycksele/Arvidsjaur	Danish Air Transport/ATR 42-72	2,100
	AB Värmlandsflyg/ATR 42	3,750
	Norwegian Air Shuttle/Fokker 50	5,650
	Nordic East Airlink/MD 82-81	5,600
	European Exec. Express/ATR 42	6,900
Arlanda–Sveg	Danish Air Transport/Beech 1900	1,000
	European Exec. Express/Jetstream 31-32	900
Luleå–Pajala	European Exec. Express/ Jetstream 31-32	1,250

*1€ = SEK 9.4

Source: Rikstrafiken.

In an attempt to address overall transport policy aims Rikstrafiken weighted requested compensation against a few parameters relating to quality of service. These were pricing and tariff system, traffic solution, accessibility for disabled persons and quality of service. The model used for the evaluation lacked transparency and the evaluation was carried out in an arbitrary manner. However, the quality of service parameters did not change the outcome of the tender as indicated by the requested compensation above.

Among acceptable tenders, DAT submitted the most advantageous bids for six of the ten routes. Wermlandsflyg acquired one combined route and European Executive Express acquired two routes. Since Regulation (EEC) 2408/92 stipulated that a two-month period needed to elapse before any selection is made, winning bidders did not receive notice until August. This gave them a mere two-month period to prepare for operations to start at the end of October 2002.

On the six routes operated by DAT loud criticism was heard immediately over perceived operational shortcomings and inabilities. DAT allegedly operated in breach of virtually all PSO requirements. Criticism spilled over on to Rikstrafiken and reached proportions such that DAT and Rikstrafiken effectively agreed that the best solution was that the contract be terminated. DAT gave notice of termination of the contract in December 2002 and ceased operation in June 2003, just seven months after commencing.

In March 2003, Rikstrafiken contracted two air carriers – Swedline and Skyways – to make an emergency take-over of the six "DAT routes". These contracts were valid only for a period of six months, a time frame estimated by Rikstrafiken to be sufficient for a new tender procedure. Because of the short contract period, requested compensation rose by more than 300 per cent compared to the figures shown in Table 7.2 for these six routes. Total annual compensation rose to an all-time high in 2004 of approximately €15 million (see Table 7.3). Despite a new tender exercise for the remainder of the first three-year round, requested compensation remained at the same high level.

The Cabinet decided in May of 2003 that the PSO routes were more important than the other transport services funded by Rikstrafiken and so was instructed to prioritise. No additional allotment of funding was made and the proposed imposition of PSO on five additional routes was halted.

The number of passengers on all PSO routes was 148,900 in 2004. Average annual compensation accumulated for all routes was €13,491,000, leaving the subsidy per passenger at around €90.

Lessons Learned from the First Contract Round

The first contract round was turbulent and resulted in poor output, heavy criticism and immense cost increases. Rikstrafiken identified the following main obstacles:

Staffing The first tender procedure was administered by a few people who lacked experience and sufficient know-how.

Lead times Lead times were underestimated, for example, the time for the government to decide on a PSO, the time necessary for translation and publication of documents in the EU OJ and the time necessary for a new air carrier to commence operations without prior experience of the route.

Tender evaluation model The evaluation model used was not consistent and allowed for arbitrary selection on the part of Rikstrafiken. Hence it lacked transparency and did not reflect Rikstrafiken's preferences in a correct manner. This in turn made it difficult for the agency to validate information submitted in the tenders. The conclusion was that a poor drafting of the tender invitation documents will inevitably lead to poor tenders from the air carriers in return.

Table 7.3 Conditions on PSO routes at the end of the first contract round

Route **No. of pax, 2004**	**Compensation accrued (€)/yr**	**Air carrier**	**No. of roundtrips according to PSO**			**Aircraft**
			Mon–Fri	Sat	Sun	
Arvidsjaur–Lycksele–Arlanda 53,000	3,250,000	Skyways Express	2	1	1	Fokker 50 (50 seat)
Gällivare–Arlanda 39,000	3,049,000	Swedline	2	1	1	SAAB 2000 (50 seat)
Hemavan–Arlanda 7,000	1,481,000	Swedline	1	0	1	Fokker 50 (50 seat)
Pajala–Luleå 4,000	1,200,000	European	2	0	1	Jetstream 31 (19 seat)
Vilhelmina–Storuman–Arlanda 28,000	2,193,000	Skyways Express	2	1	1	Fokker 50 (50 seat)
Sveg–Arlanda 5,200	775,000	European	2	0	0	Jetstream 31 (19 seat)
Torsby–Hagfors–Arlanda 4,800	610,000	Swedline	2	0	0	Jetstream 31 (19 seat)
Östersund–Umeå 7,900	933,000	Nordic Airways	2	0	0	SAAB 340 (33 seat)

Co-operation with the regions When Rikstrafiken entered the scene the affected regions were not invited to participate, which made them suspicious of Rikstrafiken's steps and led to misdirected and misinterpreted information.

Contractual incentives If an air carrier failed to comply with stipulated PSO requirements there was little Rikstrafiken could do. The penalty injunctions stipulated in the contracts were designed so that any injunction decided upon by Rikstrafiken had to be reinvested in the air service, this being an idea stemming from the dawn of the agency. It proved difficult to find suitable and meaningful ways to reinvest fundings once a failure had occurred, and in practice such reinvestment had to be agreed with the air carrier itself. Hence, the whole arrangement did not provide the incentives that were intended.

Rikstrafiken's Second Contract Round: A New Mindset

To offset the shortcomings from the first contract round Rikstrafiken decided on the following goals for the second contract round 2005–2008:

- Costs must be significantly reduced.
- Contracts shall be signed with financially stable air carriers.
- Output and quality of service shall be retained.

A project team was commissioned to work in a much more process-oriented way with a team-leader working exclusively on the tender procedure and all related documentation. The project team comprised representatives from the regions, the aviation authorities and independent expertise.

The PSO requirements largely remained the same as indicated in Table 7.1, with smaller adaptations to seat capacity supply per annum. The requirement on seat capacity per single trip was removed in order to open up to several types of aircraft. A time schedule with all necessary steps and procedures was adopted early in the process, with all lead times well kept in mind. The aim was to allow air carriers two months to submit their tenders and at least six months for winning carriers to prepare for the start of operations. The evaluation model made use of criteria specially adapted to these routes and relations. The criteria were divided into two main categories: economic organisation and technical organisation. Evaluation criteria were made mathematically consistent and stipulated beforehand. Thus the arbitrary model from the first round was abandoned. The basic qualification criteria were designed so that only carriers with a valid Operating Licence and an Air Operators Certificate for the aircraft type to be used were eligible for a contract. In addition, a record of non-payment regarding taxes or social fees during a period of two years counting back from the tender expiry date resulted in disqualification.

Rikstrafiken received over 50 bids from nine different air carriers, by then the most competitive outcome ever. Though, soon after the tender result and the signing of new contracts in the spring of 2005 new challenges suddenly emerged.

Notification to Pursue Commercial Operations and the Cancellation of a Contract

Rikstrafiken had left the opportunity open for any air carrier to notify commercial operations even after disclosure of the tender result and the signing of contracts. According to the procedures a commercial notification regarding a specific route had to be submitted not later than two months before the start of operations. In the event of an acceptable notification, the tender procedure and any contract signed regarding that particular route was to be terminated.

One carrier – European Executive Express – had failed to comply with the basic requirements in the tender and had therefore been disqualified. The air carrier then made use of the possibility to notify commercial services. Rikstrafiken chose not to accept the notification because it lacked credibility. It was submitted by the very air carrier that had failed to comply with the basic qualification criteria for a contract, and the financial status of the carrier was poor. The dispute was settled in administrative court, the issue as such being unprecedented and thus important from a principle point of view. It turned out – according to the court ruling – that Rikstrafiken had no legal right to stop an air carrier from pursuing commercial operations on a PSO route. The circumstances at hand – that the air carrier had submitted a failing bid and was on the brink of bankruptcy – did not matter. Rikstrafiken had no alternative but to accept the notification and to annul the contract on the route (Sveg–Arlanda) in question.

As the first contract round expired on the Sveg–Arlanda route at the end of October 2005, the air carrier European Executive Express commenced commercial operations following its notification to do so. A month and a half later the air carrier ceased operations and went bankrupt. Rikstrafiken resolved the problem by allowing the two competing air carriers that actually met all conditions in the original tender to submit revised bids. This led to a new contract which was then in force all through the second contract round.

Figure 7.5 shows percentage load factor by route in 2007 and Figure 7.6 the number of passengers carried on each route that year.

The Transfer of a Contract from One Air Carrier to Another

Inspired by European Executive Express's successful commercial notification on the Sveg–Arlanda route, another air carrier – Swedline Express – submitted a notification claiming it would operate the PSO route Gällivare–Arlanda on a commercial basis. Swedline had also participated in the tender procedure regarding the very same route but it requested approximately four times as much subvention as the winning air carrier – Skyways. With the court ruling mentioned above, it was clear Swedline was going to have its way with the notification and hence Skyways was not going to operate the route on a contractual basis.

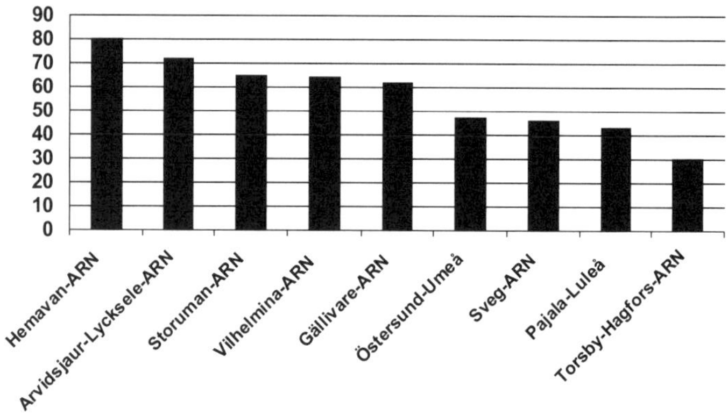

Figure 7.5 Actual load factor (%) by route in 2007

Source: Rikstrafiken. Rikstrafiken measures the true load factor on combined routes like Arvidsjaur–Lycksele–Arlanda by simply adding passenger volumes and dividing it by the total number of seats. This is possible where the capacity is shared continuously. Hemavan, Storuman and Vilhelmina share capacity to varying degrees depending on seasonal variations in demand. Therefore true load factor on these routes has varied between 50 and 90 per cent, with average levels presented above. Note: ARN = Arlanda airport in Stockholm.

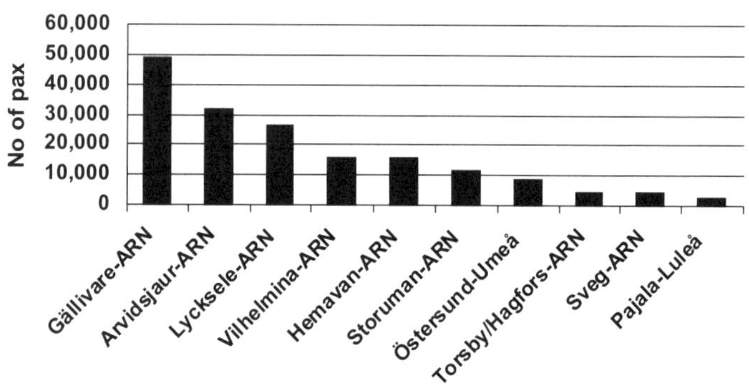

Figure 7.6 Number of passengers by route, 2007

Source: Rikstrafiken. 2007 saw the largest number of passengers on the PSO routes since 2001, i.e before Rikstrafiken entered the scene. This has happened in conjunction with something of an economic boom in the regions concerned. Indeed, at a seminar held by Rikstrafiken in 2007 the local government commissioner in the remote municipality of Pajala claimed that Pajala enjoys something that resembles an economic boom for the first time since the Korean War. Part of the reason, though, is likely to be attributed to viable air services during the entire contract round. Note: ARN = Arlanda airport in Stockholm.

Unlike European, Swedline met the basic conditions enabling the carrier to enter into a contract with Rikstrafiken. This opened up the possibility of a transfer of the contract from Skyways to Swedline, all other things being equal. In September 2005 both air carriers jointly agreed on the terms for a transfer, which Rikstrafiken accepted. Swedline operated the route for about six months before it faced insolvency and became subject of a reconstruction procedure. Following this, Swedline initiated a transfer of the contract. From the procedure it became clear that a financially sound air carrier could do well with the compensation stipulated under the contract and so – evidently – it was not the PSO route and the terms inherent with that operation that caused the problems. This led Rikstrafiken to accept a second transfer of the same contract – this time to Nordic Airways from June 2006.

Nordic Airways then operated the route under the same original contract until it too began to suffer financial turmoil due to a costly accident, the loss of a large client and skyrocketing fuel prices. As the contract on Gällivare–Arlanda approached expiry and a new one for the third round has already been signed for 2008–2011, a final transfer of the contract has been accepted by Rikstrafiken. NextJet – the winning tenderer for the third contract round – has agreed with Nordic Airways to enter into the original contract merely ten weeks from the termination of air services under that contract (see Table 7.4).

Table 7.4 Conditions on PSO routes at the end of the second contract round

Route No. of pax 2007	Compensation accrued (€)/yr	Air carrier	No. of roundtrips according to PSO			Aircraft
			Mon–Fri	Sat	Sun	
Arvidsjaur–Lycksele–Arlanda 58,000	1,615,000	Skyways Express	2	1	1	SAAB 2000 (50 seat), Avro RJ 100 (112 seat)
Gällivare–Arlanda 48,000	1,449,000	NextJet	2	1	1	SAAB 2000 (50 seat)
Hemavan–Arlanda 15,000	361,000	Skyways Express	1	0	1	Fokker 50 (50 seat)
Pajala–Luleå 2,400	864,000	Barents Airlink	2	0	0	Beech King Air 200 (9 seat)

Table 7.4 *Concluded*

Vilhelmina–Storuman–Arlanda 27,000	1,763,000	Skyways Express	2	0	1	Fokker 50 (50 seat)
Sveg–Arlanda 4,700	860,000	NextJet	2	0	0	Beech 1900 D (19 seat)
Torsby–Hagfors–Arlanda 4,700	800,000	NextJet	2	0	0	Beech 1900 D (19 seat)
Östersund–Umeå 8,100	1,034,000	Nordic Airways	2	0	0	SAAB 340 (33 seat)

Additional Compensation for Jet Fuel Price Increases

Almost immediately after the air carriers had submitted bids for the second contract round in the spring 2005 fuel prices began to increase rapidly, soon to reach unprecedented levels. Skyways, who had won the major contracts on the longest routes, argued that conditions were dramatically changed in a way that would harm the enterprise severely unless it was compensated.

Having analyzed the situation Rikstrafiken agreed to a model whereby additional compensation is paid when fuel prices exceed a certain level that bidders would have based their calculations for compensation on. An addendum to the contract was signed outlining the principles for fuel compensation. All four contracted air carriers were offered an addendum based on the same principles so as to avoid discrimination between them. From here on Rikstrafiken's contractual compensation includes a variable component.

The number of passengers on all the PSO routes amounted to 167,900 in 2007. The average annual compensation accumulated for all routes has been €8,746,000, including variable fuel compensation. Rikstrafiken has remunerated all air carriers every three months ex post based on average monthly prices for jet fuel. On average, annual fuel compensation has been €1,293,000, i.e. 15 per cent of the total compensation. However, this does not reflect true annual payments because of dramatic fuel price increases especially in 2008. The average subsidy per passenger amounts to approximately €50 with variations from less than €30 up to €338 (see Table 7.5). Despite the highest fuel prices ever it is quite clear that Rikstrafiken's aim to reduce costs succeeded. The average level of compensation went down from €13.5 million per year in the first contract round to €8.7 million per year in the second, a fall of 35 per cent (see Figures 7.7 and 7.8).

Table 7.5 Compensation per passenger and per passenger-km in 2007

PSO-route	Compensation €/passenger	Compensation €cent/pax-km
Arvidsjaur–ARN	21	3
Hemavan–ARN	22	3
Gällivare–ARN	25	3
Lycksele–ARN	30	5
Vilhelmina–ARN	48	9
Storuman–ARN	78	13
Östersund–Umeå	126	43
Torsby/Hagfors–ARN	166	59
Sveg–ARN	195	60
Pajala–Luleå	338	175
Average	48	7

There is a notable difference between the relatively large and lengthy routes that require less compensation and the small and short ones that require more. This pattern is further accentuated as a result of the third contract round from 2008 to 2011. Rikstrafiken has not defined any specific level of compensation that would be deemed unacceptable from an economic point of view. Therefore, as far as the level of compensation is concerned, the sky is the limit, as it were. Note: ARN = Arlanda airport in Stockholm.

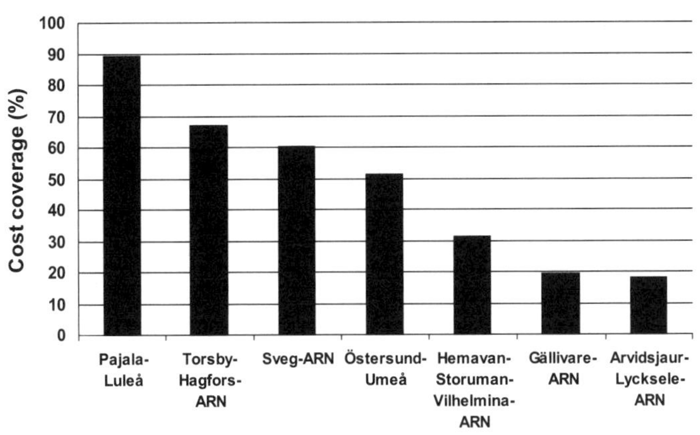

Figure 7.7 Cost coverage in comparison to total revenues per PSO route (%)

Source: Rikstrafiken. Routes with small volumes and of relatively short distances (190-320 km) depend heavily on Rikstrafiken's compensation and have little prospect of ever becoming commercially viable. On all routes combined Rikstrafiken's compensation in 2007 accounted for 31 per cent of total revenues generated from the air services. Note: ARN = Arlanda airport in Stockholm.

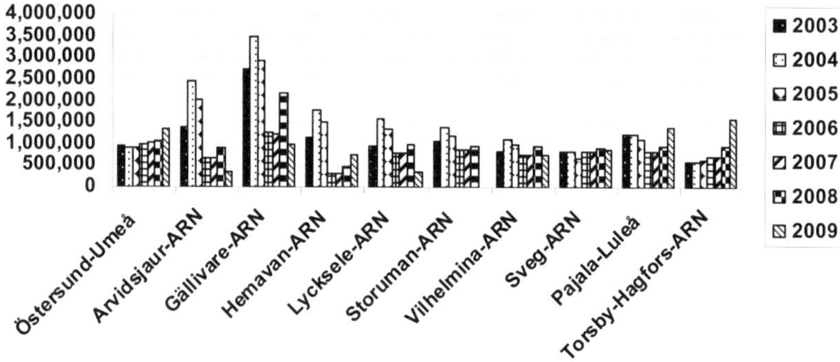

Figure 7.8 Contract costs (€) per annum by PSO route, 2003–2008

Source: Rikstrafiken. Some PSO-routes have seen considerable variations in terms of compensation requested. This is true especially for relatively large and lengthy routes. The very high costs in 2003-2005 are clearly attributed to Rikstrafiken's shortcomings in the first contract round and the fact that emergency measures had to be taken. Costs for 2008 exceed those for 2006 and 2007 on all routes which is solely due to the extreme fuel prices. Note: ARN = Arlanda airport in Stockholm.

Lessons Learned from the 2nd Round

Except for one single route the second contract round has been much more stable than the first. Nevertheless, Rikstrafiken has made the following observations to be considered in future evaluations.

Fuel prices Fuel can make up for 20 per cent of an air carrier's operational cost when crude oil exceeds USD 100 per barrel. Over time there is a strong correlation between crude oil and jet fuel prices. To Rikstrafiken, which is a government agency financed by appropriation, the development is extremely difficult to forecast. Yet, Rikstrafiken has deemed that the agency is in a better position than the air carrier to bear the financial risk associated with potentially extreme jet fuel prices.

Type of aircraft Rikstrafiken's preferences as regards the type of aircraft to be used on a given PSO route needs special attention in the tender documents. The aircraft itself is simply much of what the entire air service is all about and will have a strong impact on how passengers perceive the service and how local politicians will value Rikstrafiken's efforts. Rikstrafiken received two bids that obtained almost the same value based on the criteria stipulated. Only the type of aircraft differed and that was not a stipulated criterion. Having focused primarily on the cost issue, Rikstrafiken was on the brink of having to accept a smaller aircraft for negligible savings.

Notification to pursue commercial operations The possibility for an air carrier to notify its intention to pursue commercial services on a PSO route should not be retained after the tender expiry date. If it is, it may be detrimental to the tender procedure.

Maintaining a coherent system or splitting the routes with different contract periods An air carrier can suddenly suffer great economic or operational problems making it unable to meet PSO requirements or other contractual terms. Evidence from the first contract round shows that an emergency tender procedure to find a new air carrier for the remainder of a three-year contract round will boost the level of compensation requested. This speaks in favour of splitting the tendering system so that a new three-year round is launched if a contract has to be terminated in advance.

On the other hand, evidence from the second tender procedure shows that great savings can be made through economies of scale and scope by tendering all PSO routes simultaneously. This speaks against splitting the system and in favour of maintaining the same contract round for all routes. There is a trade-off to be made in cases like this and so far Rikstrafiken estimates that the better alternative is to maintain the same contract round for all routes. To facilitate this option, Regulation 1008/2008 should allow a Member State to take any action necessary to sustain uninterrupted air services as long as the tender procedures are correctly performed at least every four years.

Rikstrafiken's Third Contract Round 2008–2011

Reassessment of PSO

In February 2007, Rikstrafiken published a consultation report assessing the PSO routes and the merits of a continued PSO regime on those routes. Part of this report was a cost benefit analysis (CBA) never before carried out in respect of this issue. As expected, costs exceed benefits to society as a whole for most routes. In fact, on some routes the negative external effects from environmental pollution alone were estimated to equal the benefits to passengers. This means that even if one were to disregard the cost of producing the air services, benefits may not exceed the costs to society as a whole. In other words, there are PSO routes where the benefits to consumers are but a fraction of the costs associated with producing the air services. This puts a huge question mark to the issue of continuing PSO status on such routes. But then again, Rikstrafiken's objective is justified by transport policy objectives like accessibility and socio-economic development rather than economic efficiency. That is why the purpose of CBA is merely to provide guidance as regards prioritising given the limited budget.

Rikstrafiken has also calibrated a model whereby accessibility to and from a particular region can be measured. When analysing the merits of PSO status on a route the model will show whether or not a particular air service actually adds to accessibility. In the consultation report from 2007 Rikstrafiken argued that accessibility to and from Torsby, Hagfors and Storuman would not be severely affected in the absence of sustainable air services.

In the report Rikstrafiken also proposed certain amendments to the PSO. The minimum amount of single trips to be produced on each route per year was explicitly specified. Minimum seat production was adapted to recent passenger traffic development. As regards tariffs the former price-cap was abandoned and replaced by a limitation to average revenue, the idea being to gain better protection against monopoly pricing while at the same time allowing the air carrier to differentiate tariffs at their own will. In addition, Rikstrafiken proposed that the requirements on air carriers as regards financial solvency be addressed in the PSO. Following consultation in the industry the revised PSO was adopted in September 2007.

Gaining from the Experiences Made

For the third contract round a project team was formed in the same manner as for the preparation of the tender for the second round. Thanks to the systematic approach taken in the previous tender process the scene was set for a well-structured reuse of the experiences gained then and a resolution of the issues that had been identified as problematic. There were also some new issues that arose during the work.

PSO and the Content of Tender Documents

An important change was introduced in the tender documents with respect to how the material content of a PSO was made part of the tender documents. In 2005, the actual PSO text (as formally expressed in the Cabinet decision to impose them) was part of the tender documentation as a means to convey the service requirements. This time the tender documents only made reference to the fact that PSOs were imposed and then went on to express the various requirements without reference to the PSOs. There were two main reasons why this approach was taken. First, it made possible a structure of the tender documents in three parts: an invitation to tender (specifying among other things the requirements on the bidder/operating carrier), a model contract and a specification of the requirements on the carrier in preparing for and carrying out the service. The PSO text touches on all three of these parts. Second, the PSO text leaves several details unresolved that needs to be clear in the tender documents, e.g. details of the operations to be performed. Instead of "clarifying" the PSO (decided by the Cabinet) in the tender document, it seems more natural that Rikstrafiken would formulate a tender that is based on the PSO but formally is an entirely independent document, thus precluding criticism that Rikstrafiken had interpreted the PSO incorrectly. In this way it also underlined that Rikstrafiken is the government authority that is refining the formal decision to impose a PSO into a tender and subsequent contractual obligations on the part of the carriers.

Another issue during the preparation of the tender documents concerned the amount of commercial information about the routes that should be included in the tender documents. In principle it was clear to Rikstrafiken that the more information about the commercial conditions of the routes that was conveyed to the potential bidders (e.g. number of passengers, total revenues per route, annual compensation under current contracts etc.), the better bids would be received. Thanks to the reporting requirements placed on the operating carrier Rikstrafiken possesses rather detailed data of various aspects of the air services being carried out, and some of these data were publicised at an aggregate level in the above-mentioned report assessing the PSO routes. Still, to some extent it was considered a confidentiality issue to compile these data and provide them as part of the tender documents. Therefore, the final explicit information in the tender documentation was a table with passenger volume data and the amount of economic compensation from previous years for each route.

Notification of Commercial Operations after the Final Date for Tender Submission

As was experienced during the previous contract period, notification of commercial operations may be given as late as two months before the start of operations, according to Regulation 1008/2008. Such notification can in other words be given at a much later date than the final date for tender submission. This is potentially detrimental to tender competition since a carrier willing to operate the route on commercial terms has an incentive to await the outcome of the tender process, and possibly win a contract, before giving notification.

This problem was dealt with in the administrative procedures for the tender, simply by stating that the final date for giving notification of commercial operations was the same as the final date for submitting a tender and that, given a viable notification of commercial traffic, the tender procedure would be abandoned for the route(s) in question. However, whether this approach would stand a tougher test is open to question. This time there was no carrier giving notification of commercial traffic and thus no problem arose. But it is unclear what would happen if a carrier were to challenge the approach and submit a notification of commercial operation once the outcome of the tender is publicised, simply disregarding the rules given in the tender procedure.[2]

2 There is another problem in Sweden related to this. There is no government agency commissioned to supervise that carriers carrying commercial traffic on routes where PSOs have been imposed actually fulfil the obligation. Once the notification is given, Rikstrafiken and Luftfartsstyrelsen (Swedish CAA) check that the requirements are fulfilled at the time of the notification. After that point in time the carriers are on their own.

Bundling of Routes and Deciding on "Allowable" Stopovers

For the third contract period PSO status has been imposed on 11 routes. As a result of the study performed, it was decided that the tender would concern only ten routes (Storuman was eventually dropped by the politically appointed board of Rikstrafiken, but services to Torsby/Hagfors were still to be procured). Historically, air services on several of these routes had been tendered for and carried out in a combined way, i.e. starting at A, stopping over at B and proceeding to Stockholm (and vice versa). However, the rationale behind these combined routes can be questioned. Although some combinations would clearly not be viable for geographic reasons, it was also clear that viable combinations of routes beside the historic ones could not be ruled out. It would be quite possible to let the combination of routes be part of the tender process, and the bundling of routes a result of the tender process rather than an input to it.

The basic questions here are two. First, who is best fitted to decide what is economically most efficient – the community of carriers participating in the tender process with their respective bids or the public servants at Rikstrafiken? The answer to this question is by no means self-evident. Clearly, Rikstrafiken has over the years gained an overview of the air services, the demand from the passengers and their willingness to pay for the services provided to them. Second, leaving the question of combination of routes open, depending on the bids, may give some carriers (the carriers with a more in-depth knowledge of this particular market) the upper hand when bidding, which may have adverse effects on competition in bidding for the actual provision of the service. In addition, Rikstrafiken has to consider the risk that a winning bid is based on a route setup that at the end of the day would not be accepted by the passenger community.

Closely associated with this issue is the question of allowable stopovers. Only one of the 11 PSOs prohibits stopovers, and one allows double stopovers. But the PSOs say nothing about *where* the allowable stopovers may take place. Obviously it needed to be clarified in the tender where stopovers were allowed. Rikstrafiken was thus faced with the following four options: allowing stopovers only at airports where a PSO had been imposed for the route to Stockholm and which were part of the tender; allowing a stopover at Storuman (the single airport for which a PSO was imposed but which was not part of the tender); allowing stopovers at other airports lacking commercial operations at the time of the tender; or, finally, allowing stopovers at (smaller) airports with commercial traffic at the time of the tender.

The bundling issue was resolved (in the Swedish manner) through a compromise. In addition to the basic routes, bids could be placed on five predefined combined routes. Three of these combinations of routes were those with stopovers at locations that were part of the tender and that were being operated in that manner at the time of the tender (as a result of the previous tender). Two of them however were innovations in the sense that they *could* represent viable routes depending on the result of the tender; a (winning) bid on these two routes could change the final route definitions. As it turned out, bids were only placed for the three routes that already were being operated as a result of the previous tender. This does not necessarily mean that the other two routes were not economically viable per se; one must remember that changing route setups between contract periods involves cost, and a new alternative route setup must therefore be substantially better than the old one to justify a change.

As for allowable stopovers it is principally uncontroversial to allow stopovers at airports that are part of the tender or at airports that have a PSO imposed but which are not part of the tender (only Storuman in this tender). In such cases the question boils down to the bundling of routes addressed above, i.e. how much freedom should be given to the bidders when defining bundled routes. The other two alternatives however bring with them certain complications. Allowing stopovers at airports that lack commercial traffic could be considered as state subsidies in conflict with EC regulations. And allowing stopovers at airports that have commercial traffic could be considered an intervention in the competition on a market that in principle is free.

Consider Gällivare for example, perhaps the most obvious case, located some 100 km south-east of Kiruna (with a regular commercial service) and some 200 km north-west of Luleå (also with a regular commercial service). The PSO speaks of a route to Stockholm with one stopover permitted, but during the previous two contract periods the possibility to stop over has not been used – the service has been a direct route to Stockholm. The two obvious places for stopovers are ruled out because they have commercial traffic. However, the question is which market should we be actually considering? It seems reasonable to shift the perspective to the market for trips to Stockholm from the northern parts of Sweden (i.e. ≥ 1,200 km from Stockholm). So while we do not hesitate to create a subsidised direct route from Gällivare to Stockholm carrying close to 50,000 passengers annually, we rule out the option of creating a feeder line to Luleå or even a route to Stockholm with a Luleå stopover. From a taxpayers' point of view this seems slightly illogical. The PSO could just as well have pointed out (perhaps as one alternative amongst others) that a feeder route to either of the airports with commercial traffic could be introduced.

In the tender this issue was settled by explicitly pointing out the airports where a stopover was permitted. Out of the ten routes included in the tender, stopovers were prohibited on four routes (while the PSOs only prohibit stopovers on one of them). Out of the remaining six routes stopovers at airports with PSOs imposed were allowed on four routes and stopovers at airports without PSOs imposed (and without commercial operations at the time of the tender) were allowed on two routes.

Triangular Routing – the Legitimacy of Capacity Sharing

During the second contractual period one of the operators altered the setup of one of their routes at the beginning of the third and final year of operation under the contract. This concerned the bundled route Arvidsjaur–Lycksele–Stockholm with approx. 60,000 passengers annually. (Lycksele is located approx. 100 km south of Arvidsjaur en route to Stockholm.) At the beginning of the contract period the route was Arvidsjaur–Stockholm with a stopover in Lycksele in both directions. At the beginning of the last year the operator, without consulting Rikstrafiken, changed the setup. The carrier started picking up passengers in Lycksele bound for Stockholm when stopping over in Lycksele on the northward leg to Arvidsjaur and skipping the stopover in Lycksele when returning to Stockholm from Arvidsjaur. Similarly, the carrier picked up passengers in Stockholm bound for Lycksele and Arvidsjaur, flew non-stop to Arvidsjaur (passing Lyckelse), turned the aircraft around, made a short jump to Lycksele with passengers disembarking and embarking and returned to Stockholm.

Given a governing principle of first come first served, this is clearly a more optimal usage of the capacity at hand. Nevertheless it was deemed unsatisfactory by both the airports involved (Arvidsjaur and Lycksele) and Rikstrafiken, and the contractual relationship became slightly soured. Since the carrier was producing a service that fulfilled the PSO using the traditional route setup, and only used this "triangular" routing in operations that were in addition to the basic requirements, Rikstrafiken remained in tacit opposition but took no further action for the remainder of the contract period.

During the preparation of the tender for the third contractual period it was clear that it needed to be clarified whether such a route setup should be allowed. The local airports represented in the project team were opposing it, but had few arguments of any substance. It can be safely assumed that one of the major concerns of the local airports was the fact that the setup would reduce the usage of the airports in terms of take-offs and landings, thus reducing the associated revenues. Rikstrafiken was initially divided on the issue. It was clear that allowing such route setups could be more efficient for the regular air service at the two airports in question, but it was unclear if there was a risk that it would be frowned upon by the passengers and lead to undesirable results in customer satisfaction surveys or even in reduced ticket sales. Also, making it generally allowed to build triangular routes could have an impact on other routes that were difficult to foresee, at least without a deeper analysis that would require time not available in the tender preparation. These uncertainties were the main reason that Rikstrafiken decided that if such a setup was used, it would not merit in the same way as the conventional route setup when calculating the amount of operations (seats and roundtrips) produced by the carrier. In other words, Rikstrafiken was not prepared to pay for such route setups, but would (perhaps) accept it if the carrier could make it economically feasible on its own merits.[3]

---

3 Actually, Rikstrafiken has the ability to stop a triangular route setup on contractual grounds on the route in question. Although the PSO for Lycksele, closest to Stockholm, allows a stopover, thus making a Lycksele–Arvidsjaur–Stockholm route possible, the requirements specification for Lycksele does not allow a stopover. Taking passengers from

Combination Bidding and Bidding with Capacity Restrictions

A procedure for allowing the carriers to place combination bids and bids with capacity restrictions was introduced already in the tender for the second contract round. While in the tender for the first contract round bids could only be placed on the individual routes tendered for, it was recognised during the tender preparation for the second contract round that this left little room for the carriers to express their individual economies of scale when bidding. A carrier with abundant capacity might for example be prepared to lower its bids given that it would win several routes. In such case it should be possible to place a bid for several routes *on condition* that they are all won in the tender. Similarly a carrier with limited capacity might be able to place very competitive bids for more than one route *on condition* that it can only deliver up to a certain capacity. In tenders that involve placing bids on several contracts at the same point in time, and when bids can only be placed on each individual contract (route), such expressive power is not available to the bidders.

The procedure simply stated that bids for several contracts (combination bids) with an associated price were allowed on condition that individual bids were placed on each contract (route) that was part of the combination.[4] Furthermore the procedure stated that a bidder could place bids for more contracts than it would be able to carry out on condition that its capacity restriction was expressed in an unambiguous way. There was no limitation as to the number of combination or capacity restricted bids a carrier could place. In the tender for the second contract round 24 combination bids were placed, 19 of them from the carrier that was eventually disqualified. No capacity restricted bids were received. Although the cost of compensation to the carriers decreased by some 35 per cent, the "combination discount" was a meagre 0.3 per cent.

In the tender for the third contract round, the same procedure was used in the tender with only pedagogical and cosmetic changes. This time 13 combination bids were placed (all from the same bidder), but as in the previous tender there were no capacity restricted bids. The resulting combination discount was 8 per cent.

Lycksele to Stockholm via Arvidsjaur would therefore be in contradiction of the traffic specification. Ironically, this restriction was not introduced with the intention to stop a triangular route in this particular case, but was a result of the mindset that once en route from Lycksele to Stockholm it should not be allowed to have a stopover located geographically between Lycksele and Stockholm.

4 The purpose of the requirement for individual bids on contracts that are part of a combination is to ensure that at the end of the day there exists an allocation of contracts that cover all the contracts tendered for. Without this restriction, combination bids might overlap in a way that prevents allocation of all contracts.

Bid Evaluation Model

The qualification criteria and the specification of requirements for the preparation and provision of a service open up a multidimensional "space" of bids that are all potentially viable. The purpose of the bid evaluation model is of course to rank any bids that fall within these limits in a transparent and consistent manner, reflecting the preferences of Rikstrafiken. In the tender for the second contract round the criteria used were related to the economic and technical organisation of the carrier and the service provided, e.g. support function for "space control", support function for "operation planning", ticket distribution system and ticket price differentiation. Each criterion was rated according to clear and transparent rules and the bid price adjusted by a predefined and publicised sum, different for each route, depending on the total rating.

The type of aircraft used was however not part of the criteria and, as noted above, Rikstrafiken was near to having to accept a smaller aircraft on one of the routes for negligible savings. This experience was a clear warning sign that something was missing in the way Rikstrafiken's preferences for bids of various compositions had been expressed in the tender documents. It became obvious that Rikstrafiken needed to be able to justify the selection of the winning bidder in view of all conceivable differences between bids.

In the preparation of the tender for the third contract the approach concerning the evaluation model was changed. Some of the criteria that had been subject to evaluation in the previous tender had become obsolete due to changes to the PSOs.[5] The other criteria were changed to hard requirements that had to be fulfilled by every bid, thus raising the minimum level of quality that had to be fulfilled by the carriers.

Given that most of the old criteria were changed into hard requirements the new evaluation model took aim at reflecting the preferences of Rikstrafiken and defining aspects of possible bids that would lead Rikstrafiken to choose a more costly bid before a cheaper one. These criteria were developed in workshops together with representatives from the local airports. This exercise proved to be very fruitful. Each involved party was expected to state the net worth in costs for the conceivable quality differences between bids, taking one factor at a time. That is given two bids, equal in all but, for example, the number of seats available, how big must the cost difference (the saving) be in order to justify the selection of the bid with "less" quality?

As it turned out, four criteria were defined that together with the bid price would decide the winning bid. These criteria were cabin size (number of seats available), travel time, preparation plan and marketing plan. For each criterion it was defined how much the bid would be adjusted upwards or downwards depending on the

5 In the previous tender the PSOs stated a ceiling price for each route, and the evaluation model rewarded bids where ticket prices were below the ceiling price. This time the PSOs were changed to state maximum average ticket revenue per quarter.

fulfilment of the criterion. As for cabin size, bids would be adjusted upwards if the seat capacity of a bid for a route was below certain levels, and a simple table was created, which for each route defined how the bid sum should be adjusted (see Table 7.6). The table can be interpreted as the minimum saving Rikstrafiken wanted to see in relation to a bid with a certain size of aircraft in order to accept a smaller aircraft.[6]

Table 7.6 Tender criteria (capacity): bid adjustment with respect to cabin size (approx. €000)

No. of seats in aircraft	< 100	< 70	< 50	< 40	< 30	< 15	Maximum adjustment
Arvidsjaur–Stockholm	64	53					117
Gällivare–Stockholm	96	64					160
Hagfors–Stockholm					8	8	16
Hemavan–Stockholm				11	32		43
Lycksele–Stockholm	64	53					117
Sveg–Stockholm					16	32	48
Torsby–Stockholm					8	8	16
Vilhelmina–Stockholm							0
Pajala–Luleå					8	8	16
Östersund–Umeå							0

The travel time criterion was treated in a somewhat different fashion. The criterion stated that in the evaluation every bid would be adjusted upward with a fixed sum per minute of travel time, thus rewarding bids with less travel time.[7]

The evaluation model also covered planning documents that the bidder was required to submit with its bid. These planning documents concerned the preparation phase for starting up the air services and the marketing activities to be performed before and during the contract round; and according to the tender, the commitments made in these planning documents would be made part of the contractual obligations of the winning bidder. The criteria outlined ten issues that were to be addressed for the preparation phase and five issues that were to be addressed for the marketing activities, and stated that the content of the submitted documents would be rated on a scale of 0–10 and 0–5 respectively. The resulting

6 The requirements specification stated the minimum number of seats for most of the routes.

7 Also in this case, the requirements specification for each route stated what maximum travel time would be accepted.

score would then be multiplied by a constant, different for each route, and the resulting sum would be deducted from the bid value. In this manner the signal was sent to the bidders that an investment made in preparing for the startup of the traffic and in planning of the marketing activities would be rewarded during the bid selection by as much as €50,000 for the singularly most heavily used route (and less than €5,000 for the least heavily used one).

Outcome of the Tender

In the tender for the third contract round six carriers participated. They placed a total of 35 bids, of which 13 were combination bids all placed by the same carrier. Two of the bidders were disqualified, one due to tax debts and one due to failure to obtain its AOC within the deadline (two months after tender submission).

Nex Time Jet, which won nine of the ten PSO routes, placed all the combination bids and won the routes with one combination bid covering three "singular" routes, plus three combined routes (see Table 7.7).

Table 7.7 Amount of compensation for the third contract round (€)

Route	Air carrier	Oct 2008–Oct 2009	Oct 2009–Oct 2010	Oct 2010–Oct 2011	Total compensation
Arvidsjaur/Lycksele–Arlanda	Nex Time Jet	672,917	699,834	727,827	2,100,578
Gällivare–Arlanda	Nex Time Jet	990,149	1,029,755	1,070,945	3,090,849
Torsby/Hagfors–Arlanda	Nex Time Jet	1,538,096	1,599 620	1,663,604	4,801,320
Sveg–Arlanda	Nex Time Jet	865,179	899,786	935,777	2,700,742
Hemavan/Vilhemina–Arlanda	Nex Time Jet	1,490,030	1,549,631	1,611 617	4,651,278
Östersund–Umeå	Nex Time Jet	1,326,608	1,379,672	1,434 859	4,141,139
Pajala–Luleå	Scand. Air Ambulance	1,377,660	1,446,808	1,518 617	4,343,085
Total		7,587,722	8,605,106	8,963 246	25,156,074

Note: In addition to this amount extra compensation is forecast at approximately €2,600,000 for the entire contract round at crude oil prices of $97 per barrel (at end of September 2008). Assuming passenger volumes at the 2007 level and crude oil prices at no more than $100 per barrel, average subsidy per passenger will increase moderately from €50 to €55 during the contract round 2008–2011.

Lessons Learned So Far from the Third Contract Round

Air services under the third contract round have just commenced at the time of writing this chapter, so lessons learned will have to be restricted to reflecting on the tender procedure, the outcome of the tender and the framework within which the tender has been carried out.

Number of Carriers with Interest in the Tender Process

First of all it is worrying that the number of carriers willing to place bids has decreased. In the previous round there were nine bidders, while in the recent tender there were only six, none of which were new carriers participating in the tender. Since two of the bidders were disqualified during the bid evaluation, there were in fact only four carriers competing for the contract. It seems clear that Rikstrafiken has to be more active to encourage bidders, including from outside Sweden, to participate in the tender. Merely making the tender public according to the required procedures is not enough.

Price Structure

The price structure in the submitted bids is also of concern. The winning carrier won all routes except for one, in "splendid isolation" as for the level of requested compensation. This is a clear indicator that Rikstrafiken needs to put an effort into enhancing the competition when preparing the tender for the fourth contract round. If this is not achieved it seems there is a risk that the level of requested compensation for the next tender round will "gravitate" towards a point between the current level of compensation and the level of the (much) less competitive bids received in the recent tender round. Apart from encouraging more carriers to bid for the next contract round, the single most interesting approach to be considered in order to achieve convergence in the level of requested compensation would be to allow for iterative bidding in the tender procedure, i.e. an auction procedure.

Combination Bidding

The fact that it is possible to submit combination bids and bids with capacity restrictions could be of importance when recruiting carriers to bid. Experience from the two recent tender rounds tells Rikstrafiken that only a few carriers use the possibility of combination bidding and none use capacity restricted bids. Why this is so is not known. It could be that if the possibility of submitting combination bids and bids with capacity restrictions is communicated more effectively to the community of carriers, it would result in more carriers participating in the tender.

Economies of Scale

There appears to be economies of scale in gaining more contracts rather than just one. The winning air carrier has left a discount of approx. 8 per cent for the combination of all routes, except Pajala–Luleå, compared to the sum of requested compensation for each of these routes. This further fosters the idea that Sweden should continue striving for tender procedures where all PSO routes are handled simultaneously rather than splitting the system up using different contract periods.

Tender Marketing

Rikstrafiken depends on competition to maintain a reasonable level of compensation. To attract air carriers to these tender procedures must therefore be a continuous task and requires that Rikstrafiken participates in various domestic events where air carriers themselves participate, be it conferences, seminars and the like. Similarly, Rikstrafiken should enhance its presence in Europe to make European air carriers aware of the prospects on offer in Sweden. So far, Swedish national carriers have seen little or no competition from other European air carriers. Unfortunately, Sweden shares this experience with the rest of Europe.

Routing Methods

It seems that the issue of "triangular routing" must also be revisited. As stated above, it is clear that capacity could be better used this way, thus affecting the economic viability of the routes and the level of requested compensation. Rikstrafiken needs to decide whether triangular routing in principle can be accepted, but also, as indicated above, to carefully consider where it may be applied in order not to lead to unacceptable route setups.

The Tendering Framework

The framework within which the tender is carried out consists primarily of the decisions to impose a PSO under EU regulations. It seems that the structure and content of the PSOs can be improved. For example the capacity to be produced is expressed in several ways in the Swedish PSOs, and they are not entirely consistent. Also some details, e.g. size of aircraft to be used and rules for how the supply may be altered as a result of shifting demand, could perhaps be left out of the PSO text and instead be decided by Rikstrafiken. It seems clear that Rikstrafiken would be capable of transforming a more generally worded PSO decision into the detailed specifications and requirements that are required in the tender process.

An important question for the future is also whether it would be possible to express the formally decided PSO in a way that does not pinpoint the creation of a route between A and B, i.e. to express the PSO rather as a decision to provide for *some* suitable air traffic solution, and leave it up to Rikstrafiken to design that solution.

Conclusions

The continued work in Sweden to improve the tendering of PSOs must now focus on attracting more carriers to participate in the tender process for the fourth contract round. This will be imperative in order to maintain and if possible reduce the level of costs which was established in the second contract round and which in principle was maintained for the 3rd contract round. In order to do so several actions must be taken. The tender must be marketed more efficiently. The content of PSOs, the service requirements formulated by Rikstrafiken and the tender procedure must also be scrutinised to identify elements that may keep carriers outside Sweden or the Nordic countries from bidding. An important issue is how Rikstrafiken can provide potential bidders with hard commercial facts about the routes in order that potential bidders with less detailed market knowledge are prepared to participate in the tender. The structure of the tender documents must be reviewed to ascertain that it is similar to what carriers in other parts of Europe are familiar with.

The possibility to place combination bids should be maintained in the future and should be communicated more effectively and pedagogically to the community of carriers, as well as the possibility to place capacity restricted bids. In particular, the latter may attract smaller carriers to be more active in future tenders.

Maintaining or reducing the level of cost for the PSOs will also depend on continued work with developing route efficiency. Stereotype patterns for conducting the operations must be scrutinised to find new and more efficient ways of combining routes and carrying out the PSO air services.

Chapter 8
Tendering for and Operating PSO Routes

Basil O'Fee

Introduction

This chapter will describe Highland Airways' (HA) experiences with Public Service Obligation (PSO) tenders. The chapter is written in a subjective manner to shed light on how the PSO system has been perceived by this airline. Highland Airways has been tendering for PSO routes over the last ten years and currently operates two routes in Scotland and one in Wales.[1] Two of the routes are served by 19-seat BAe Jetstream 31 aircraft and one with a nine-seat Britten Norman 2 Islander. The experiences described here may hopefully be of relevance to others.

Context of the Tendering Process

In Europe, the European Union (EU) has created a legal framework by which governments can subsidise air services. In attempts to create a free and open market the EU wanted to stop national governments subsidising their national airlines. However it was recognised that, in certain circumstances, air services did require protection or subsidy and the Public Service Obligation was born.

Each national government seems to adapt and use the PSO 'tool' slightly differently. In the UK historically it was only used to underwrite 'lifeline' services. In other countries, notably Ireland and France, PSOs have apparently been used to underpin regional development and link the regions very affordably with the capital.

In the UK there have been attempts, so far unsuccessful, to ring fence slots into busy London airports like Heathrow and Gatwick using PSO criteria. Some regions have lost, or might lose, daily services to these airports as slots are increasingly sought by international routes. The PSO would most likely not require subsidy.

Hitrans (the Highlands of Scotland regional transport forum) recently proposed an ambitious proliferation of PSO routes in Scotland with the aim to improve connectivity and travel affordability in the region. The aim was to stimulate traffic to such an extent that subsidy would eventually not be required. One insight seemed to be that networks needed to be optimised instead of using just point-to-point routes (which seems to be a PSO legislation presumption). This proposal appeared as too interventionist for the taste of the UK and Scottish governments.

---

1 Highland Airways ceased trading in March 2010, some time after this chapter was written. The text therefore reflects the situation at the time of writing.

The Scottish government launched the less ambitious Air Discount Scheme (subsidising remote community residents with discounted travel) in response to that wave of agitation.

The recent Welsh PSO was a new departure for the UK in that a PSO linking north and south Wales was not justified under purely lifeline criteria such as access to hospitals and metropolitan services. It was understood as strengthening Welsh national integration and the role of Cardiff as the Principality's capital. It is notable that such an idea should have arisen from a Welsh Assembly, and apparently never crossed the mind of any previous UK-centred administration.

Each country needs to get Brussels to approve their PSO justification. This can be a long process before approval and ultimately the issue of tender.

The Tendering Specification

A PSO tender usually boils down to some minimum service criteria, and a request for the amount of deficit funding required to supply such a service. The specified criteria typically revolve around frequency of flights, number of seats or size of aircraft and a stated maximum permissible fare.

Very quickly an airline can work out which aircraft will be ideal for the job, and which aircraft will be most cost effective. Some jurisdictions seem to be more generous or demanding than others. In Ireland and by repute Norway and France we observed that aircraft tended to be specified as larger than absolutely necessary for the anticipated passenger numbers. In Scotland we noticed that the minimum aircraft size and price was usually sought.

Curiously, these tenders rarely mention freight capacity in their criteria, even though that may often be a key element of the service – for lifeline mail and newspapers coming in or for priority parcels and perishable produce (such as live seafood) coming out of the remote community.

The Response

When bidding for a completely new PSO route one has no historical data to help predict passenger numbers. In Wales the Welsh Assembly Government's consultant had indicated that 9,500 passengers would use the new PSO route. Highland Airways were rather wary of this figure, especially as we learned that the consultant had compared Anglesey's relationship with Cardiff with that of Campbeltown's with Glasgow. Glasgow was the nearest metropolitan area to Campbeltown and hence it seemed there would be much more administrative, health-related, higher education, shopping, friends and family and even leisure travel generated. There was no other practical option for Campbeltown when its inhabitants sought metropolitan services.

Anglesey had a completely different relationship with Cardiff. Its nearest cities are Liverpool, Chester and Manchester. This PSO was being fostered to confront generations of separation between the north-west and south-east of Wales. Indeed it was claimed that it was a conscious effort at 'nation-building'. What was going to happen with such a new service? Local sceptics poured scorn on the service, claiming it was a service aimed at the rich and wealthy. Ultimately nobody knew and it was the airline's money potentially at stake.

Out of prudence we assumed 7,500 would travel in the first year, 8,500 in the second and 9,500 in the third. In fact the first year was a resounding success and our fears proved completely unfounded: 14,133 passengers travelled in the first year, which has only just completed as this chapter is being written, with 80 per cent+ load factors. Nonetheless as the operator we could not take too big a risk. If the numbers had not showed up, no one was going to bail us out or have much sympathy with our three-year contract potentially acting like a millstone.

In established PSOs with good historical passenger data, like the Western Isles route between Stornoway and Benbecula, the forecasting risk is much reduced. Passenger numbers have been within 1 per cent or 2 per cent of the preceding year, over the period of seven years of our involvement. In Oban and the Inner Hebrides there was no historical data and numbers on the islands served are very small. However it is too soon to say whether we have over- or underestimated the numbers travelling. Figure 8.1 shows Highland Airways' PSO routes.

Figure 8.1 Highland Airways' PSO routes

Costs

The main costs that cause an airline concern are the cost of crew, the cost of handling, airport charges and taxes, the cost of ownership and insurance of the aircraft, contingency costs for breakdowns, passenger liability costs and the cost of fuel. We will now discuss some of these cost elements in turn.

Aircrew

Being a small airline we tend to exist near the bottom of the feeding chain, or career path, for most pilots. Some pilots do make a full career with the likes of us, and some do return, but many of our pilots perceive us as a stepping stone to larger or VIP aircraft. Thus the pay rates for crew are really set by forces outside our control. The challenge in recruiting and training pilots has become one of the key dilemmas for our sort of air operation. We have several ways to counteract this tendency.

We have our own cadet programme for young aspiring pilots – taking them from Private Pilot's Licence level through to their Commercial Pilot's Licence, and then ideally retaining their services for a minimum of five years. This offering is heavily oversubscribed, as we give a pilot a way to qualify without developing a mountain of debt.

We have also been forced to trawl the world for pilots and tend to run through different countries as friends follow friends through our ranks. New Zealanders, Australians, South Africans, Europeans and Canadians are regularly represented in our aircrew ranks. Despite all this we consider that our recruitment standards remain demanding and our training is of a high calibre. Our pilot 'product' is therefore all the more attractive to the larger airlines. They can often outbid us with more glamorous equipment, easier working conditions and enhanced pay.

Nevertheless, pilots also understand that working with us allows them to stay close to their original love of flying. They get the chance to fly with minimal navigation aids in the north of Scotland, in small aircraft, in beautiful surroundings, and they undertake diverse aerial tasks. Many leave with a heavy heart, knowing that their most enjoyable days of flying now probably lie behind them, in their time with us. This is a true Faustian pact for them.

Airport Handling

In the Western Isles we decided to arrange our own handling with our own staff. We were unsure if the existing handlers would treat us equitably as we had won the contract from their airline employer.

Running our own handling allowed us to retain more control on the softer sides of the interface with our customers, and ensured that we were able to create our own identity for our services. In fact we believe that we have been successful in developing an air service that retains the informality, friendliness and warmth that is possible on such small aircraft and in remote communities.

In the Western Isles, our handling staff were also able to service some of our other contracts. They helped unload the newspapers and mail aircraft in the morning. They converted the newspaper aircraft into a passenger aircraft for the PSO by removing the freight bags used to carry the papers. They also saw off the mail in the afternoon and looked after the fisheries patrol aircraft whenever it passed through Stornoway. A small additional income stream was serving some of the General Aviation traffic that came through Stornoway, sometimes from the other side of the Atlantic.

Most readers of this book will never have heard of Highland Airways. In the Western Isles our name loomed large with all arriving and departing flights. Our joke was we were 'The West's favourite airline'.

Airport Charges and Taxes

This section reflects our view that the airports often exploit their position as 'spatial monopolists'. It sometimes seems that the airlines take all risks and the airports take the cream off the top. When they are setting their charge rates they do so, in our opinion, without the same amount of soul-searching or self-sacrifice as airlines regularly practise. According to this author's experience, the only airport that was really under the cosh of close scrutiny was Dublin, where Michael O'Leary of Ryanair would not let it easily pass on its charges to his passengers.

But there are positive experiences as well. We had a very pleasant experience with Cardiff Airport when launching the new Welsh PSO. The airport manager and his team well appreciated the political priority for the route, and they supported us above and beyond what could be justified by the numbers of passengers involved. Their Media Relations and Marketing department worked very closely with us and made sure we found our way around the Welsh media much more effectively.

Highland and Islands Airport was also very generous with its press officer, and we released many joint press releases over the years. Our experience was largely positive with all the airports with whom we worked. Unfortunately in most PSOs, the local authorities seem to want the ticket price to include all the extras such as airport charges. They think this is more transparent to the passengers. We disagree on this. The passenger should know what a large proportion of their fare feeds the airport through which they pass. We also find good reasons to ask why the airline should collect charges for the airports. If the passenger had to pay the airport charge at the airport entrance, or at security, there would be more public awareness and possibly more downward pressure on the cost regimes of these 'spatial monopolies'. However we are not holding our breath for such a change.

This all-inclusive price approach also has the result that when airports put up their charges, inevitably out of synchronisation with our PSO contract, our yield is nibbled away. We can only guess on their future price profile when preparing our tender. One property manager for an airport we dealt with understood this reality and set our rents to be reviewed at the time of the re-tender, which was helpful.

Fortunately, our PSO aircraft fall below the size that qualifies for the UK's Air Passenger Duty (APD). This tax indulgence makes our smaller aircraft more competitive and attractive for PSOs as a result. Currently the APD is £10 per sector. In the Highlands and Islands of Scotland there is a partial APD cease in recognition of the higher number of flights that a regional inhabitant needs to make to reach their final destination. However in Wales this indulgence for smaller aircraft avoids an extra £20 tax take on the circumscribed £100 return fare.

There is talk of extending the reach and increasing the amount of aviation tax, which could hit PSOs hard. We have lobbied against this intended change. However the additional deficit incurred by operating a PSO will no doubt bear an uncanny resemblance to the additional duty. It will be a matter of what the government takes away with one hand it will have to pay out with the other.

Contingency Costs

What happens when the aircraft breaks down on one of our PSOs? How often is this going to happen, and how expensive is it going to be to deliver a reasonable response? These are the sorts of calculations an airline must make when preparing its bid.

With an Inverness base supporting the Western Isles we have been able to deliver excellent and cost effective support options whenever we had technical or crew illness problems. Usually we had been able to hurry out other Inverness-based aircraft to pick up the pieces when we had problems. Our record of service robustness was excellent and it was five years before we had to cancel a service because we had no other aircraft to cover for a technical problem.

The Cardiff-based aircraft proved to be a different challenge, and in retrospect we badly underestimated how much it would cost when things went wrong. This was further complicated because RAF Valley (Anglesey) found it difficult to grant evening extensions because of resource issues. The result was that we only had about 20 minutes to sort out problems on the afternoon rotation before we started to have to consider cancellations or diversion to Hawarden (which is near Chester and quite a road journey from Anglesey).

We soon felt the heat of displeasure whenever we had to 'road' passengers because of technical failure. Our first failure unleashed a torrent of outrage from the unlucky passengers. Nobody seemed to humbly accept the fact that before the PSO this was what they had to endure as a matter of course. Transferring replacement aircraft and crew between Cardiff and Inverness proved to be an expensive business. But once historical data exists, all these calculations become more accurate.

One should bear in mind that pioneering PSOs or any new entrant airlines on a PSO endure a higher risk, as they have to make estimates of these sorts of intangible costs. We proved to be seriously adrift with our contingency factor in Wales. Fortunately this additional cost was counterbalanced by other positives.

Operating a PSO

As a small niche operator, we made the decision to attempt to turn our small size to our advantage. We would position ourselves as the community airline, with local staff handling, local accents and even local languages (e.g. Welsh or Gaelic) being used to get closer to our customers.

In Highland Airways each major customer has a primary contract manager who orchestrates our company's relationship with the client. Their aim is to maximise business with the customer and to ensure healthy and happy relations. We have been able to give sufficient attention to the sponsoring authority so they understand that their PSO business is important in our company. We believe that a larger airline might struggle to maintain this depth of relationship. Our contract manager also takes an interest in the micro-marketing required to operate small PSOs.

Micro-marketing

Achieving press coverage is a very cost effective method of raising the profile of the service. This may be maximised by our PR representative providing national and local media outlets with information. We also offer free or discounted tickets to selected journalists covering stories to these communities. We ensure we have a bank of pictures, stories, facts and figures on the service so that journalists can easily obtain and incorporate information into their articles. We ensure we identify publicity opportunities within radio, TV and film production.

We take advantage of the fact that most PSO destinations are remote and well loved by the nation of which they are part. This affection can be translated into an ease in capturing the public's imagination for a story about the place. Coverage can be out of all proportion to the passenger numbers involved. Who wants to read about a new Cardiff to Newcastle service for instance? Yet Cardiff to Anglesey received much press interest. It was relatively easy to also generate stories about the Outer or Inner Hebrides.

Collaborations

Forming collaborations with selected partners has been the best way for the promotion of PSO services especially because of our limited budget and thin manpower. Other stakeholders, we anticipate, should include such organisations as:

- the local sponsoring authority and the local enterprise company promoting business in the area;

- airports that can be useful in promoting the service;
- the tourism authority, hoteliers and other tourism interests;
- ferry, car hire and bus operators;
- business and freight users of the service;
- ATOL[2] tour organisers and local travel agents;
- local newspapers and radio stations.

It is appreciated that increasing numbers of travellers are now using the internet to explore possibilities and plan their itinerary. We place stress on ensuring our website is attractive and up to date.

Highland Airways uses an industry standard internet booking system. This system is used by other small to medium-sized airlines and was the most cost effective one we were offered for our small passenger numbers. It has a yield management capability, with a low-cost airlines approach. This permits quotes for the travel cost to be varied by the airline based upon the database's knowledge of numbers of seats booked, time to flight, target yield and other 'rules' that the operator can specify. The system also produces useful sales and marketing statistics and reports that can better guide future pricing decisions.

We had to devote considerable effort to becoming adept in this technology. Before we adopted the system we relied on a call centre and paper and pencil system for all our bookings. It worked reasonably well. We knew the cost invested in the technology would not deliver a requisite increase of passengers in our captive PSO markets. However we also knew that we *had* to make the migration into this technology.

Pricing and Special Offers

Highland Airways has certain maximum fares specified by the PSO sponsoring authority. We also used a set of special incentive fares to increase numbers, especially in predictably quiet periods. Additionally we are keen to keep some simplicity in the system so that travellers develop a sense of the range of fares to expect. The route manager decided to offer some strongly discounted seats on the various PSOs that rewarded advance bookings on all HA flights. The logic was that this was potentially creating 'new' travellers and revenue rather than recycling and discounting travellers who would have to fly in any case. We believe this practice lifted occupancy levels despite our average yield per seat being reduced as a result.

Ambiguously the sponsoring authority inevitably wishes us to have these discounted fares for populist reasons. On the other hand we know that if our yield is too depleted by such measures, our tender will fail to be competitive. When members of the public complain that they missed the headline or discounted fare, we remind them that in a PSO *all* the fares are discounted!

2 Air Travel Organiser's Licensing.

Advertising

We engage in formal advertising only sparingly at PSO launch times, or in the run up to special events such as Christmas. We always felt our resource was not adequate for any sustained advertising campaign and in small communities the news got around quickly enough in any case.

We always point out to the other stakeholders that we were only three-year-tenants (the length of the PSO), and any above target income will be taken back by the local authority. As a consequence we have no incentive, apart from professional pride, to spend money to deliver above target yield.

As a result, we encourage the local authority and the airports concerned to invest in the routes, as they 'own' the routes in a much more real sense than we do. We have had increasing success in communicating this perspective, as we have better understood it ourselves, to the other stakeholders over the years.

Launch Week

The launch of a completely new passenger service attracts good media coverage (in both Wales and Oban) and we take as much advantage of this as possible. Press flights, launch events and various media interviews are all part of pushing the message out. In both Wales and Scotland we feel these completely new services benefited from very good local media coverage.

In our tender submissions we tend to skew our marketing spend towards the beginning of the contract, where it will have the longest effect on our three year tenure. We might spend on something like a 4:2:1 ratio over the three year period.

The Community Airline

In new PSO routes we are always touched when local activists who have lobbied hard for such a development over many years see their ambitions fulfilled. Their sheer joy at seeing the service begin is tangible. They are more aware than we of how many years effort was required. As an example, we clearly remember the pride that public representatives of Anglesey took when the 'Anglesey' destination was visible for the first time on departure screens and above check-ins at Cardiff Airport. A similar sentiment was expressed by Councillor Duncan McIntyre of Oban when the Inner Hebrides service was being launched. In fact he said his hair turned white during his long efforts to bring about the desired outcome over many years of lobbying and effort.

At the user end of the spectrum on touchdown for the first time on the island of Coll much of the community came out to welcome the new aircraft. The whole primary school was there, delighted to see the new airport finally enter service. The smiles of joy and the sense of history was also strong among the adults who came to welcome the new service.

A Trilingual Website

Since the service launch in Wales we have been working with the Welsh Language Board to develop our language policy for the service in line with local legislation (which is more demanding than in Scotland). They wanted us to develop a bilingual version of our website and offered support to do so. However we also operated PSO services in Scottish Gaelic-speaking areas, and we knew that if we only offered a Welsh version of the site, voices would soon be raised in our own backyard in Scotland.

Therefore in 2007 we decided to present a trilingual website, offering users English, Welsh and Scottish Gaelic options. Apart from some Swiss regions, we believe HA is the only business in the world to offer this unusual mix of languages. Initially, there were over 90,000 words involved on 300 different web pages. Translators and programmers worked hard alongside our webmaster to make this happen, and on a very tight budget. Our route manager persuaded us to work with VisitWales to deliver a local call centre solution based in Bangor, which also offered a welcome bilingual service. We know that this local flavour helped reassure users that the service was offered on a human and very accessible scale.

All these initiatives to be frank, we knew, would deliver us little extra direct business. However the judgement was taken that our principals would appreciate these sorts of features and our travelling customers would look on us more kindly as a result. We recognised that we did not operate in a purely market-driven environment.

Cabin Crew

Cabin crew are not legally required on HA's small aircraft, and in the Hebrides our flight crew deliver the safety brief. This requirement ensures that the flight crew interact more closely with passengers and indeed get to know the regulars. In many ways this was a positive experience for both crew and passengers.

The Welsh Assembly Government however encouraged us to use cabin crew on our small Jetstream 31 aircraft. We can say that their efforts and care greatly enhanced the travel experience of the passengers. We were fortunate to have recruited a dedicated and charming team.

Local Flavour

The following copy of the speech that this author gave at the press launch of HA's new PSO service in Anglesey in April 2007 will give a flavour of the different issues and sentiments at work in these PSOs.

> We at Highland Airways have only come recently to this process of providing an airlink between North and South Wales.

We appreciate that this new service represents the vision, the ambition and the practical efforts of many people, both in the north and south of Wales. We are impressed with this new air terminal and look forward to working with Operon and Babcock. We are impressed with the helpful attitude of the many organisations and agencies that we have already had dealings with (the Welsh Assembly Government, the RAF, Anglesey Council, Cardiff Airport, VisitWales, University of Wales in Bangor to mention a few). We are also impressed with the many good wishes we have received from those who have every intention of using this service. On a personal level I am gratified to observe a spring in the step that I sense in modern Wales.

We have no real idea whether this service is going to build slowly or quickly. We understand that many hope that this service represents a first step in improving the connectivity of Wales and also the links between north Wales and the wider world.

However we do know that we have the ideal aircraft type and size for this route at this stage in the process. We also know that we have experience of running comparable services in the north of Scotland. We hope that our efforts and experience will strengthen this venture.

Our team has now joined those collective efforts of which I spoke. Our business development manager and route development manager are working hard to promote the service in collaboration with others. Many of you will have met him. Our operations manager is looking to the ground handling arrangements and crew rostering. Our chief pilot and our CEO are ensuring we have an adequate supply of trained and experienced pilots. Our engineering manager is working to ensure that the aircraft is technically reliable and well maintained. We intend having at least two flight crew and two cabin crew based in Cardiff as a result of this new service. Our quality manager is independently assessing all our arrangements and preparations with an eye to safety and quality levels of service delivery.

The map of Wales is changing with this service – not the physical map (that is provided for us by the immutable efforts of the creator) – but the mental map. No longer will the diagonals of Wales seem so very far apart. Let us all – locals and visitors to Wales – take this new fact to our hearts as we consider our Welsh business travel and leisure travel options.

Small Communities, Strong Voices

Of course it would be idealistic to believe that such significant community initiatives and investments as new PSOs are passed and supported without local controversy. A whole raft of objections can be expected, and the airline inevitably can be drawn into the maelstrom. Objections can be simplified to:

- There are better things to spend this money on.
- This is only a public sector fat-cat service designed to make life simpler for our lazy public representatives – no one else will use it.
- Aviation is environmentally damaging, and we should not be subsidising it.
- The selected solution has terrible shortcomings that undermine its promise – frequency, aircraft type, timetable or expense.
- I fear change.

One should have in mind that any new PSO proposition will run into objections not only from government finance departments but also, more surprisingly, from some of the members of the community they are designed to serve.

We have also noticed how one or two unreconstructed objectors can harass public officials relentlessly in the media and on the letters pages of the local press. Small communities can tend to be buffeted by such forceful personalities. As an airline we strive not to get involved, although sometimes in their anger objectors also try to blacken our service as additional support for their views. Inevitably, the media mischievously delights in such discord and also wants to give airtime to contrarian views, such as from the environmentalists.

One distinction we did recognise was that these remote communities are not big polluters. It is hard for this author to recognise that metropolitan sophisticates who have developed a green perspective would deny these fragile communities lifestyle enhancements because of the depredations of mankind elsewhere.

Sustainability, we realised, should also weigh in the computation of pros and cons. The very viability of these PSO-supported communities, many of which are struggling to survive economically and demographically, is in question. Keeping these communities vibrant and attractive to child-rearing couples and young ambitious school leavers seems absolutely critical.

Our colleague and route manager on the Wales service had an interesting experience:

> On 9th May 2008 I was taking part in a radio phone-in show for BBC Radio Wales, celebrating the first anniversary of the Intra-Wales Air Service. Other guests included Martin Evans from Wales Transport Research Centre and Neil Crumpton from Friends of the Earth Cymru (FOE).
>
> I had been warned that Neil Crumpton was a seasoned campaigner against the air service. There was some concern that we may get a hard time from Friends of the Earth, as they had been campaigning against the air service on a number of fronts. They were against the use of a larger aircraft, didn't want to see any increase in frequency and were very concerned about the potential of mass

tourism destroying Anglesey. The presenter was primed to provoke fierce debate between his guests and callers.

At reception I happened to bump into Neil Crumpton whilst waiting to be escorted through to the studio. I decided to take the opportunity to make small talk whilst waiting, and didn't hold back on telling him all about the service. It transpired that lots of assumptions had been made by FOE, based mainly on information inaccurately gleaned from media reports about issues such as aircraft size and engine type. I found that in chatting with Neil I was able to brief him properly on what we had achieved, share some information on numbers of passengers carried and offer realistic projections for the future. Neil admitted that he had been misinformed and that this information entirely changed his view on the service.

Later we went on air and when the presenter tried to get an argument going it fell rather flat when Neil admitted that the FOE position had changed and that they had no objections to the service based on the information provided by me. *But despite the turnaround in opinion I still couldn't persuade Neil to use the air service!*

As I explained it to myself when pondering such community discord over the years – small communities can have big arguments.

Reflections and Recommendations from an Airline's Point of View

The PSO Legislators

The legislation should be changed to create an incentive for operators to sell more seats than budgeted. Currently, 100 per cent of any above budget yield is returned to the sponsoring authority. We think a 50/50 share of success would give the airline some incentive to try to move the PSO route out of subsidy. There is after all this ultimate aim to nurture the PSO to survive without subsidy, but this will rarely happen with the airline so counter-incentivised.

One should ensure all tenders have a lead time that permits real competition. Perhaps this should be enshrined in the legislation. At the moment tardy or overworked public procurement departments can leave things to the very last minute, ensuring that only the incumbent can succeed.

The Sponsoring Authority

A good tender should ideally have a pre-tender briefing (either collective or individual), allowing the sponsoring authority to explain their priorities and selection criteria. The tender should be run with a suitable period for response

– four weeks seems to be the current favoured period, and it is surprising how many tenders come out a few days before a Christmas or Easter break, ensuring our tender team has to sacrifice their holiday.

The preparatory and set-up period should be generous and adequate. Public authorities appear to believe that crew and aircraft are lying around awaiting their call, and that four or six weeks is an adequate set-up period. In these days, six weeks hardly provides enough time to obtain security passes from airports for new staff. Crews need to be advertised for and interviewed (six weeks); have to give notice at their previous jobs (one to three months depending upon seniority); they need potentially to be converted to type and find housing – maybe move their family; and they need to be inducted and given some line training (one to two months). Ideally the set-up period should be three to six months depending upon the complexity or exoticism of the aircraft and route.

Authorities do not seem to realise that breakneck preparations lead to more expensive aircraft, staff and supplier solutions that will eventually feed back to affect the size of subsidy. We have taken to itemising a one-off set-up charge, as we are tired of paying for temporary accommodation and other exceptional costs incurred because of pressured preparations.

The sponsoring authority should engage an experienced aviation consultant to guide and help them in their selection, if the procurement department does not have the expertise. This consultant can also track any agreed Key Performance Indicators (KPIs) for the operator (punctuality, reliability etc.) during the running of the service. Periodic contract reviews are good in keeping all parties focused and collaborative.

Local Councillors and other Stakeholders

The sponsoring authority sets the criteria for the PSO, and activists should ensure that these criteria suit the communities they are being designed to serve. The maximum fares set in the PSO deserve some scrutiny and reflection. What is the logic for these fares? Are they equitable and comparable across other PSOs or across Europe? What is the rationale for these figures? In our experience there is little rationale apart from best guesses. Different countries have come to different conclusions about affordability and social inclusion.

We believe that the community should also reflect on the freight needs for the PSO, and perhaps express this in the tender criteria. For instance, Campbeltown is a small town, and one of its main attractions is a world-famous golf course at nearby Machrihanish. Often the international golfers are keen to catch the flight and yet their clubs may sometimes not be able to accompany them because of aircraft size. Barra has a PSO to its famous beach airport. However the island has a large fish-processing facility. Currently the factory cannot export the more lucrative 'live' seafood by air because of aircraft limitations. On the PSO service between Benbecula and Stornoway the Royal Mail used to put the outgoing mail on HA flights. Often we off-loaded some mail because it exceeded our guaranteed reserved quota for them. On the islands this caused much anguish and discord, but the sponsoring authority would not let us sacrifice a passenger seat or two to cater better for the mail. Ultimately, the Royal Mail had to change the routing of their dedicated mail aircraft to better cover this issue, and the revenue was lost to the PSO.

In all these cases the government could have included freight capacity in their specification. This does not happen in any PSO that we have come across. Freight rates and freight capacity could be taken into account. Size of aircraft, frequency, days of operation and timetable are all variables that greatly influence the amount of subsidy required, so the specification is crucial.

The Airline

Airlines should remind the other stakeholders of their responsibilities to the PSO and seek their collaboration, particularly in marketing. They should also earn the respect of the community through delivering punctual and reliable service and addressing any problems that do emerge with vigour.

They should show flexibility and humanity to customers and be on their side versus the endless legislation and encumbrances that aviation has foisted upon us all. We do for instance encourage our pilots to use their judgement and discretion in waiting for connecting aircraft, or waiting for late passengers. These are small aircraft in remote locations with few other travel choices, and a missed flight can cause serious inconvenience. This author can remember, for instance, when one of our pilots phoned and asked the other passengers if they minded waiting 30 minutes for an incoming flight with five schoolchildren who were connecting with our PSO flight. They all agreed, and I supported this sort of approach.

PART C
Infrastructure Provision

Chapter 9
Airport Infrastructure in Europe's Remoter Regions

Rodney Fewings

Introduction

This chapter examines the provision of airside infrastructure for a selection of countries located within continental Europe. The selected mix of countries has been deliberate so as to compare the provision of infrastructure between those countries located in the centre of Europe with those countries located on the peripherals of Europe, the hypothesis being that the level and provision of specific types of airside infrastructure will vary according to the needs of remote regions within specific countries. Remoteness may be qualified in a number of ways. Typically, remoteness might be considered as the distance from the capital city or major areas of urbanisation. Alternatively, remoteness might be linked to a lack of alternative transport modes that serve the peripheral regions of an individual country. For example, travel options between Kirkenes and Oslo are possible by either surface transport (car) or by air. The car journey takes 30 hours, while the flight time is no more than 3 hours. However, travel between Iceland, the Azores, Ireland and mainland Europe is only possible using air transport or limited ferry services.

There is, of course, a high propensity for travel to and from remote regions. Surface or air transport satisfies a need for access to education and medical facilities, and to the major business and industrial centres of a country. In the reverse direction, transport modes facilitate inbound tourism into the remote regions and hence provide a boost to the local economy. There is a consensus that the provision of adequate transport and communications infrastructure is a necessary investment to avoid the depopulation of remote areas and, indeed, that limited movement of the indigenous population from congested urban to peripheral areas should be encouraged.

The provision of adequate transport infrastructure does, however, have a price. Considerable expenditure has been incurred in a number of countries to improve the national road system to serve the peripheral regions. Current examples include the upgraded road network radiating from Dublin and serving the west of Ireland. In Norway, the road network has seen the construction of new roads and tunnels to serve communities that were previously only adequately served by shipping or aviation services. In most remote regions, rail networks are limited or non-existent. Shipping is an option but only on well-defined routes linking islands with the mainland (Ireland) or with other islands (Madeira) or as a link serving isolated coastal communities (Norway). For the purpose of this chapter, the well-used shipping links between Ireland, the UK and mainland Europe are noted but not considered further.

It is therefore not surprising that aviation has played a major role in providing a transport service to the peripheral regions of Europe. As mentioned above, there is a price to pay. The provision of a small airport, for a remote community, requires substantial capital investment and incurs significant annual operating costs. In most cases, the operating costs are not covered by revenues from passengers, cargo and other users. The provision of such airports therefore becomes a financial burden to the state or local community.

This chapter will therefore look at a number of cost-related aspects of airport provision, including the distribution of airports in terms of land area and population, runway length, opening hours, air traffic control, navigation aids and rescue and fire fighting services. As mentioned before, the countries selected range from those with extensive peripheral geographic areas to those that are part of mainland Europe so that a comparison between non-peripheral and peripheral region countries might be made.

For example, the following physical geographic summaries have been made for each country:

- **Austria**: Small land area, located on mainland Europe with good national road and rail links. Winters may be moderate to severe with extensive snow coverage.
- **Denmark**: Small land area, located between mainland Europe and Scandinavia; consists of several islands and a peninsula, with good intra-regional road, rail and shipping links. Winters are generally moderate.
- **Finland**: Large land area stretching to the Arctic Circle; limited road and rail links serving the peripheral regions. Winters are generally severe with extensive snow coverage and long hours of darkness.
- **France**: Large land area, located on mainland Europe with good national road and rail links; Corsica dependent on air and shipping services. Winters are generally moderate with snow coverage on high ground only.
- **Germany**: Large land area, located on mainland Europe with good national road and rail links. Winters may be severe with extensive snow coverage.
- **Greece**: Small land area located on the edge of mainland Europe; islands dependent on air and shipping services. Winters generally moderate with snow coverage confined to northern part of country.
- **Iceland**: An island, small land area dependent on air services to mainland Europe; limited national road network. Winters severe, with extensive snow coverage and long hours of darkness.
- **Ireland (Republic of)**: An island, small land area dependent on air services to mainland Europe; good national road network. Winters are generally mild with limited snow coverage.

- **Italy**: Large land area, located on mainland Europe with good national road and rail links; Sardinia and Sicily dependent on air and shipping services. Winters are generally more severe in the north, milder in the south.
- **Norway**: Large land area stretching to the Arctic Circle and beyond to the Russian border; limited road and rail links serving peripheral regions. Winters severe, with extensive snow coverage and long hours of darkness.
- **Portugal**: Small land area, located on edge of mainland Europe; Madeira and the Azores dependent on air services as shipping options are limited. Winters are generally mild.
- **Scotland**: Medium land area dependent on air services to outer islands; generally a good national road but rail network limited outside of central lowlands. Winters are generally moderate with limited snow coverage and long hours of darkness in the north.
- **Spain**: Large land area, located on mainland Europe with good national road and rail links; Canaries and Balearics dependent on air and shipping services. Winters are generally moderate with snow coverage restricted to high altitudes.
- **Sweden**: Large land area stretching to the Arctic Circle; limited road and rail links serving peripheral regions. Winters severe, with extensive snow coverage and long hours of darkness.
- **United Kingdom**: Island, medium land area dependent on air services to continental Europe and outer islands; generally a good national road and rail network (see Scotland). Winters are generally moderate with limited snow coverage.

  (Note that for comparison purposes, Scotland has been considered both as a separate entity and also as part of the United Kingdom.)

Lastly, the criterion for airport selection is on the basis of 2,000 commercial/scheduled passengers per annum. This means that airports at the bottom of the scale may only have perhaps two or three scheduled services per week. However, the same airports could have substantial general aviation traffic but this has not been taken into consideration. From past data analysis, the number of airports with more than 2,000 passengers per annum in any one country has increased over the last ten years although, from one year to another, some airports may drop out of each country list, only to be replaced by others.

Airport Provision in Terms of Population and Land Area

Table 9.1 shows, for each of the selected countries, the number of airports per million inhabitants. The highest ratios are found in Iceland (38.71) and Norway (10.76); the lowest ratios are to be found in the larger countries of continental Europe and in the United Kingdom. Note that in the case of the latter, the ratio for Scotland is significantly different. Iceland has a small population scattered over a relatively large geographic area and, taking into account the climatic conditions especially during winter, it is perhaps not surprising that the ratio of airports to population is very high. Norway, in second place, also has similar problems of scattered population, a large geographic area and extreme winter conditions. In contrast, the countries of continental Europe have a more even distribution of population, better road and rail communications linking the country and significantly less severe winters.

Table 9.2 takes an alternative approach by looking at the number of airports per land area (100,000km^2). The resulting ratios place the sample countries into three distinct groups. The highest ratio includes Denmark (27.91), Greece (28.79), Scotland (21.52) and the UK (22.86). The one area of commonality between these four countries appears to be the number of islands that make up the associated physical geography of each country. However, in the case of Denmark, extensive surface transport links now link the major islands together; and for the UK/Scotland the results are influenced by air transport provision to the Western Isles, Orkney, Shetland, the Isle of Man and the Channel Islands. (Note that the Isle of Man and the Channel Islands are Crown dependencies but are included with the United Kingdom for completeness.)

Table 9.1 Airports per million inhabitants

Country	Number of airports	Population (m)	Number of airports per million inhabitants
Austria	6	8.17	0.73
Denmark	12	5.37	2.42
Finland	21	5.30	3.96
France	78	60.77	1.28
Germany	46	82.22	0.56
Greece	38	10.65	3.56
Iceland	12	0.31	38.71
Ireland	13	4.23	3.07
Italy	41	59.72	0.69
Norway	51	4.74	10.76
Portugal	14	10.08	1.39
Scotland	17	5.20	3.26
Spain	42	45.06	0.93
Sweden	41	9.08	4.52
United Kingdom	56	60.59	0.92

Sources: Shown in References.

Table 9.2 Airports per 100,000km²

Country	Number of airports	Land area (100,000km²)	Number of airports per 100,000km²
Austria	6	0.84	7.14
Denmark	12	0.43	27.91
Finland	21	3.38	6.21
France	78	5.47	14.26
Germany	46	3.57	12.89
Greece	38	1.32	28.79
Iceland	12	1.03	11.65
Ireland	13	0.70	18.57
Italy	41	3.01	13.62
Norway	51	3.24	15.74
Portugal	14	0.92	15.22
Scotland	17	0.79	21.52
Spain	42	5.05	8.32
Sweden	41	4.50	9,11
UK	56	2.45	22.86

Sources: Shown in References.

A surprising omission in this group is Portugal (15.22), partly because of the low number of airports on the mainland. However, Portugal is included in the second group, which also includes France (14.26), Germany (12.89), Iceland (11.65), Ireland (18.57), Italy (13.62) and Norway (15.74). The most common factor is the geographic feature of a single land mass with few off-shore operations to islands. Note, however, that Ireland is not too dissimilar to the first group, in part due to the Aran Islands services.

The last group includes Austria (7.14), Finland (6.21), Spain (8.32) and Sweden (9.11). The latter three countries have relatively large land masses, with the population being concentrated in specific areas rather than evenly distributed throughout the entire country. Spain includes airports serving the Balearics and the Canaries but much of central Spain has very few airports apart from the Madrid system.

Airport Runway Lengths

Airports were divided into three groups: the first including those where the longest runway is greater than 1,799 metres in length, the second where the main runway is between 1,000m and 1,799m in length and the third where the runway length is less than 1,000m. The breakdown by runway length reflects the landing and take-off requirements of the main types of aircraft that operate to regional airports, namely: small turboprop equipment requiring up to a 1,000-metre runway length; narrow-bodied jets, seating up to 180 passengers, requiring up to a 1,800-metre runway length; and larger aircraft requiring a runway length of 1,800 metres or more. It is, of course, acknowledged that the precise cut-off points between individual categories are to some degree subjective.

Table 9.3 summarises the data for each of the sample countries. It can be seen that one-third of the airports in Norway have a maximum runway length of less than 1,000 metres. In almost all cases this is due to geographic terrain and obstacle restrictions, with a consequent knock-on effect on the type of aircraft and range of services that are able to operate from that airport. The same constraints apply to short-runway airports in other countries, for example on the Irish and Greek islands.

Table 9.3 Airports and maximum runway length

Country	Runway length			Total number of airports
	1,800m and above	1,000–1,799m	Less than 1,000m	
Austria	6	–	–	6
Denmark	8	4	–	12
Finland	21	–	–	21
France	59	17	2	78
Germany	32	13	1	46
Greece	26	6	6	38
Iceland	4	6	2	12
Ireland	5	4	4	13
Italy	34	6	1	41
Norway	21	13	17	51
Portugal	8	5	1	14
Scotland	9	6	2	17
Spain	37	5	–	42
Sweden	32	9	–	41
UK	35	16	5	56

Sources: Shown in References.

Airport Opening Hours

The purpose of Table 9.4 is to summarise, as simply as possible, the opening hours of the airports from the sample countries. While 24-hour opening speaks for itself, the classification of those airports open for less than 24 hours had to be simplified out of necessity. In general, airports open for less than 24 hours a day can be classified into two groups: first, those airports that open from early morning and close in the late evening. Their opening hours are consistently in excess of 12 hours a day. The second group includes airports with opening hours that are subject to many different constraints and conditions. These can include daylight opening only and those opening only for the arrival and departure of a limited number of services. These airports are often open for considerably less than 12 hours a day. Note that opening hours are those based on those quoted for air traffic services, fuelling, meteorological services and not on airport administration opening hours, which tend to be shorter. Opening hours do, of course, have important cost implications and these will be alluded to again later in the chapter.

Table 9.4 Number of airports open for 24 hours, > 12 hours and < 12 hours

Country	Opening hours			Total number of airports
	24 hours	≥ 12 hours	< 12 hours	
Austria	1	5	–	6
Denmark	5	4	3	12
Finland	6	6	9	21
France	21	36	21	78
Germany	19	20	7	46
Greece	7	–	31	38
Iceland	3	–	9	12
Ireland	3	4	6	13
Italy	28	11	2	41
Norway	9	36	6	51
Portugal	10	–	4	14
Scotland	4	7	6	17
Spain	12	26	4	42
Sweden	4	27	10	41
UK	20	24	12	56

Sources: Shown in References.

From the data shown in Table 9.4, apart from the fact that Finland, Greece, Iceland and Ireland have proportionally the highest number of airports that are open for less than 12 hours a day, it is difficult to discern any clear pattern. In terms of overall opening hours throughout the country, Italy and Portugal have the highest proportion of airports open for 24 hours. In general, 24-hour opening is generally applicable (subject to curfews) to the main international and charter airports (with the exception of Greece for the latter), whilst the opening hours of regional and local airports tend to match the timings of scheduled flights or be restricted due to lack of navigation aids for night operations.

Airports and ATC Service Provision

Table 9.5 shows the level of air traffic control (ATC) service provision for airports in each country. As there are many types of ATC service the data has been simplified as far as possible to represent the different types of service that are available at each country's airports. There are, of course, cost implications in that the provision of air traffic control services represents not only a capital investment cost (primarily the control tower and associated navigation system infrastructure) but also an operating cost that is a function of airport opening times (minimum requirements for manning levels and shift patterns).

Table 9.5 Air traffic control (ATC) services

Country	ATC services				Total number of airports
	APP	TWR	AFIS	None	
Austria	6	–	–	–	6
Denmark	2	6	4	–	12
Finland	5	10	6	–	21
France	48	18	8	4	78
Germany	18	18	10	–	46
Greece	14	22	2	–	38
Iceland	2	2	8	–	12
Ireland	3	7	–	3	13
Italy	33	1	7	–	41
Norway	8	11	32	–	51
Portugal	12	1	–	1	14
Scotland	11	1	5	–	17
Spain	39	2	–	1	42
Sweden	5	23	13	–	41
UK	47	4	5	–	56

Sources: Shown in References.

The following simplified categories of ATC service provision have been used:

- **APP** (approach) represents those airports where a separate approach and departure control service is available for that particular airport. Approach and tower (TWR) facilities are generally on the same premises (for example, both within the control tower). Approach and TWR use separate frequencies. This is the top-level form of ATC service available at most major airports. This category also includes those airports where an approach control service is provided by another nearby airport or area control centre.
- **TWR** represents those airports (usually small regional or local airports) where the same controller is responsible for providing APP and TWR control service to an aircraft. By definition, therefore, the approach and tower control facilities are co-located in the same premises and use the same frequency.
- **AFIS** (Aeronautical Flight Information Service) is sometimes provided at smaller airports if there are no licensed air traffic controllers validated to provide a TWR control service.
- Lastly, some very small airports, perhaps only used once or twice a day, have no air traffic control service available at all, although there may be an airport manager present who is empowered to switch on lighting and provide a limited amount of information to the flight crew when on the ground.

The spread of data in Table 9.5 tends to correlate with the number of major airports and traffic levels. The main point of interest is that Iceland and Norway appear to have the highest proportion of airports with AFIS when compared to the total number of airports in each country. This can be justified by the limited number of aircraft movements at many of the airports in these countries. A number of other countries (Austria, Ireland, Portugal and Spain) make no use at all of AFIS. Within the United Kingdom, AFIS airports are to be found exclusively in Scotland.

Airports and Navigation Aids

Table 9.6 shows the types of navigation aids currently installed for each country. Again, the data has been grouped into a number of simple categories, as follows:

- **ILS**: At least one runway with an Instrument Landing System (ILS) available for one or both landing directions. No distinction is made between the different levels of ILS available (Cat I, II, III etc.). ILS installation includes a combination of glide path (GP) and localiser (LLZ) equipment. At most airports with ILS, other ground aids such as VHF omnidirectional range (VOR) and distance measuring equipment (DME) are also available.

- **LLZ**: No full ILS service is available, but a localiser (LLZ) is provided which gives guidance in one alignment only, glide path equipment being not installed. At many airports with LLZ, other ground aids such as precision approach path indicator (PAPI), VOR and DME are also available.
- **Other**: No ILS/LLZ equipment is available, but at least one other navigation aid (these include VOR, DME, NDB, L, VORTAC, and TACAN)[1] is located at, or adjacent to, the airport.
- **None**: These airports appear to have no navigation facilities within a reasonable distance of the airport itself.

Norway appears to be unique in that a significant proportion of its airports are only equipped with localisers (LLZ) rather than an instrument landing system (ILS). Apart from Norway, with few exceptions, the localiser option as a navigation aid is rarely used. It should be noted that a significant percentage of airports in Scotland and Greece rely on en-route navigation aids (VOR, DME etc.), or have no navigation aids at all. With navigation aids, there is a trade-off between the capital investment and maintenance costs, and the advantage of being able to keep the airport open during hours of darkness and low visibility conditions.

Table 9.6 Navigation aid facilities by country

Country	Navigation aids				Total number of airports
	ILS	LLZ	Other	None	
Austria	6	–	–	–	6
Denmark	10	1	1	–	12
Finland	21	–	–	–	21
France	69	2	–	7	78
Germany	35	1	5	5	46
Greece	3	–	35	–	38
Iceland	2	3	4	3	12
Ireland	7	1	1	4	13
Italy	32	4	3	2	41
Norway	22	26	3	–	51
Portugal	6	–	4	4	14
Scotland	9	–	7	1	17
Spain	35	–	7	–	42
Sweden	38	3	–	–	41
UK	43	0	11	2	56

Sources: Shown in References.

1 NDB – non-directional beacon; VORTAC – Very High Frequency Omni-Directional Radio Range Tactical Air Navigation Aid; TACAN – Tactical Air Navigation.

Airport Rescue and Fire Fighting Service (RFFS)

Table 9.7 shows the level of rescue and fire fighting service provision for the countries surveyed (with the exception of Greece, for which there was incomplete data available). RFFS is a major cost investment for airports, the investment cost being linked to the largest aircraft type that uses the airport, even on an infrequent basis. The operating cost is linked both to the RFFS category and airport operating hours. For example, an airport open for 24 hours would require a two or three shift system; an airport open for less than 12 hours a day could operate on a single shift system. In many cases, airports offer a higher category of RFFS cover provided that sufficient notice is given. In most cases this means drafting in additional staff rather than providing additional fire tenders, water and foam. In recent years, the removal of remission factors has seen the RFFS category increased at many airports, usually by a factor of +1.

Table 9.7 Rescue and fire fighting service (RFFS) provision

Country	RFFS category				Total number of airports
	8 > 10	6, 7	3 > 5	≤ 2	
Austria	6	–	–	–	6
Denmark	1	5	5	1	12
Finland	4	15	2	–	21
France	13	34	23	8	78
Germany	21	12	5	8	46
Iceland	1	1	9	1	12
Ireland	2	2	4	5	13
Italy	21	14	4	2	41
Norway	2	20	29	–	51
Portugal	4	4	5	1	14
Scotland	4	6	6	1	17
Spain	15	18	8	1	42
Sweden	8	19	13	1	41
UK	24	16	13	3	56

Sources: Shown in References.

As previously mentioned, a number of simple categories have been adopted for the data, as follows:

- RFFS Categories 8 and 9: At these airports, RFFS service provision meets the requirements for wide-bodied medium/long-haul aircraft, including the A330, A340, B757, B767 and B747. Note that the A380 and B747-Stretched will require airports to meet International Civil Aviation Organization (ICAO) Category 10 for RFFS provision.
- RFFS Categories 6 and 7: At these airports, RFFS service provision meets the requirements for narrow-bodied short-haul jet aircraft, including regional jets, B737 and A320 families. Several of the airports are able to provide additional RFFS cover on request.
- RFFS Categories 3, 4 and 5: At these airports RFFS service provision meets the requirements for short-haul turboprop operations, including the Fokker 50 and ATR42/72, Twin Otter and Shorts 360. Several of the airports are able to provide additional RFFS cover on request.
- RFFS Category 2 and below: At these airports RFFS service provision meets the requirements for short-haul short runway turboprop operations, including BN Islander. The category also includes airports for which no information was available and which may be served mainly by general aviation. Several of the airports are able to provide additional RFFS cover on request.

From Table 9.7 it can be seen that RFFS provision closely follows the level of commercial airline operations within individual countries. France, Germany and Ireland have the largest number of airports with a low level of RFFS provision (two or less), while France, Germany, Italy, Spain and the United Kingdom have a large number of airports with RFFS provision for long-haul aircraft (Categories 8 to 10). France and Norway have the highest number of airports with RFFS Categories 3, 4 and 5.

Concluding Remarks

There has been an increase in the number of airports with more that 2,000 passengers per annum in recent years. This increase has in part resulted from the commercialisation of what were previously military airfields and partly from the growing influence of low-cost scheduled airlines, some of which have been making use of remotely located and therefore very under-utilised airports. The airlines have also provided passengers with a cheaper and quicker air transport alternative to surface transport modes than was previously available. Norway's high dependence on air services for domestic travel is apparent, with the country continuing to have a domestic journey trip rate per capita significantly higher than most other European countries.

In terms of airport opening hours a majority of airports in Greece (31 out of 38) and Iceland (9 out of 12) had opening hours for less than 12 hours a day on average. However, in most cases, actual opening hours are linked to the operation of scheduled services and there may be times when actual opening hours are longer than those indicated.

In terms of ATC provision at airports, very few airports have no ATC service at all. There is a significant use of AFIS in Norway and Iceland and less so in Denmark, Finland, Scotland and Sweden. Also, there are a small number of airports in France, Germany, Iceland, Ireland and Portugal with no navigation aids either at or in the vicinity of the airports. In all such cases, commercial air services will only operate under daylight and good visibility conditions.

The provision of rescue and fire fighting services has seen a trend towards an increased provision of vehicles and staff in order to meet a higher RFFS category than was previously the case. In part this is because of the removal of remission factors and this has had an impact on smaller airports. At the larger gateway airports the advent of the A380 and growth in long-haul routes has also been a driver for an increase in the aerodrome RFFS category.

References

AENA (2007), *Trafico de Pasajeros en los Aeropuertos Espanoles*, Spain.

AENA (2009), *Aeronautical Information Publication*, Spain.

Air Traffic Safety Electronic Engineers Association, *Greek Airports*, Greece.

ANA Aeroportos de Portugal (2007), *Statistics*, Portugal.

ANAM SA (2007), *Aeroportos da Madeira*, Portugal.

Arbeitsgemeinschaft Deutscher Verkehrsflughäfen (2007), *Monatsstatistik*, Germany.

Austro Control (2009), *Aeronautical Information Publication*, Austria.

AVINOR (2007), *Traffic Statistics*, Norway.

AVINOR (2009), *Aeronautical Information Publication*, Norway.

AVINOR/Cranfield University (2005), *Air Transport Dependence in European Countries*, Norway.

Bundeswehr Air Traffic Services Office (2009), *Aeronautical Information Publication*, Germany.

Central Statistics Office (2007), *Passenger Movement by Air and Airport Pairings Data*, Ireland.

Civil Aviation Administration (2009), *Aeronautical Information Publication*, Denmark.

Civil Aviation Administration (2009), *Aeronautical Information Publication*, Iceland.

Civil Aviation Authority (2007), *Traffic Statistics*, United Kingdom.

Civil Aviation Authority (2009), *Aeronautical Information Publication*, United Kingdom.

Direction des Transports Aeriens (2007), *Bulletin Statistique Trafic Commercial*, France.

Direction Générale de l'Aviation Civile (2006), *Activité des Aéroports Français: Service des Bases Aériennes*, France.

Direction Générale de l'Aviation Civile (2009), *Aeronautical Information Publication*, France.

Ente Nazionale per l'Aviazione Civil (2006), *Annuario Statistico*, Italy.

Ente Nazionale per l'Aviazione Civil (2009), *Aeronautical Information Publication*, Italy.

Eurostat (2008), *Total Population – Country Profiles*, Brussels.

FINAVIA (2007), *Finnish Airports and Airport and Destination Facts*, Finland.

FINAVIA (2009), *Aeronautical Information Publication*, Finland.

Irish Aviation Authority (2009), *Aeronautical Information Publication*, Ireland.

Luftfartsverket (2007), *Annual Statistics*, Sweden.

Nav Portugal E.P.E. (2009), *Aeronautical Information Publication*, Portugal.

Norwegian Civil Aviation Administration/Cranfield University (1998), *A Comparative Study of European Airport Provision*, Norway.

Official Statistics of Sweden (2007), *Statistics*, Sweden.

Statens Luftfartsvaesen (2007), *Annual Statistics*, Denmark.

Statistics Austria (2007), *Transport – Aviation*, Austria.

Statistics Iceland (2007), *Tourism Transport and IT*, Iceland.

Swedish Civil Aviation Authority (2009), *Aeronautical Information Publication*, Sweden.

Williams, G., Fewings, R. and Fuglum, K. (2007), 'Airport Provision and Air Transport Dependence in European Countries', *Journal of Airport Management*, 1, 4, pp. 398–412.

World Aeronautical Database, <http://worldaerodata.com>.

Chapter 10
Problems of Complying with ICAO Annex 14 at Norwegian Aerodromes

Are Lien

Introduction

In Annex 14 to the Convention on International Civil Aviation under "Applicability" it is said that:

> The interpretation of some of the specifications in the Annex expressly requires the exercising of discretion, the taking of a decision or the performance of a function by the appropriate authority. In other specifications, the expression appropriate authority does not actually appear although its inclusion is implied. In both cases, the responsibility for whatever determination or action is necessary shall rest with the State having jurisdiction over the aerodrome.

Until the year 2000 Luftfartsverket, Norway, was both the regulator and operator of most Norwegian airports. In 2000 the regulator and operator were separated. The Norwegian Civil Aviation Authority (NCAA) was established as the regulator and Avinor AS (Avinor) was established as a state-owned company responsible for operating most of the Norwegian aerodromes – 46 altogether. The separation was followed by an increased focus on the Norwegian regulation concerning the physical design of aerodromes. Work on a new standard and a national regulation had been going on for some years. An upgrade of the aerodromes was also considered due inter alia to follow up the requirements for scheduled passenger traffic with larger aircraft, especially at the short field aerodromes.

Problems of complying with Annex 14 is usually related to the physical dimensions of the runway – the strip and Runway End Safety Areas (RESA) – and the need for obstacle-free zones for safe approaches, landings and take-offs. In Norway, as a typical remote and mountainous region, we mainly face problems with complying with Annex 14 due to a lack of obstacle-free areas around the runways and in the aircraft approach and departure sectors. This chapter will focus on the development of the Norwegian regulation (BSL E) based on Annex 14 from 2000, and how Avinor is implementing the new regulation at 46 airports during the re-certification of the aerodromes in the period 2004–2009. The main focus will be on the safety areas and the obstacle limitation surfaces and the problems associated with meeting the requirements in these areas. The Norwegian regulation states that the target level of safety shall be maintained by implementing mitigation when not complying with the regulation.

Development of the (New) Norwegian Regulation (BSL E)

The Situation Before the New Regulation

The first draft of the Norwegian regulation concerning the physical design of aerodromes was developed by Luftfartsverket in 1994/95. The regulation and practices were based upon the requirements in International Civil Aviation Organization (ICAO) Annex 14 – international standards and recommendations. The standards that Norway had ratified through the Chicago convention and most of the recommendations in Annex 14 were treated as standards in the first draft of the national regulation. However, until the year 2000, Norway had also notified ICAO of certain differences between the national regulation and the international standards in Annex 14.

The width of the runway strip area was generally less than the standard used by ICAO for precision approach runways. Norway was not practising Runway End Safety Areas (RESA) on code number 1 and 2 instrument runways. All runways with code number 1 and 2 were certified as non-instrument runways even though we had published several non-precision instrument approaches. The minimum descent height (MDH) at these aerodromes was not to be established below 500ft above AD elevation without permission from the NCAA, because visual approaches were considered safer than instrument approaches.

Compared to the ICAO standards, the national requirements concerning obstacle limitation surfaces were reduced. Norway also practised certain differences from the ICAO concerning design of threshold lights, length of approach lights and obstacle lights. The main reasons for these differences were mainly due to difficulties with terrain around the airports and the cost related to following the standards and recommendations in Annex 14.

The New Regulation Developed by the NCAA

The first draft edition of the national regulation was published by the NCAA in 2000. In this draft edition all the standards and recommendations in ICAO Annex 14 were applied as standards, and in some areas the regulation was more stringent than Annex 14. Most of the Norwegian aerodromes were not able to comply with all the standards and recommendations in ICAO Annex 14, and Avinor did not accept the draft regulation developed by the NCAA. However the first official edition of the national regulation was published by the NCAA in December 2000. In this edition, the NCAA had agreed that some requirements were not in force as long as further developments and alterations were not made at the airports. It was also stated by the Ministry of Transport and Communications that these requirements should not be applied at the airports at this stage. This practice was valid until the end of 2001.

Due to the rather negative signals from Avinor concerning the use of all recommendations in Annex 14 as national standards, the NCAA was asked by the Ministry to submit a special analysis to clarify the consequences for the Norwegian airports (both financial and infrastructure). A special task force was appointed to clarify which parts of ICAO Annex 14 could be used as national regulations, and which parts were to be studied further. The task force agreed upon which recommendations to give further study/analysis, and a contract with AEA Technology plc was signed in January 2000.

The following recommendations were analysed by AEA Technology plc:

- length and reference code of runways and taxiways
- runway and taxiway strips
- the dimensions and design of runway end safety areas
- separation distances between runways and taxiways
- definition of obstacle limitation surfaces.

These recommendations were considered to be the major problems to comply with at most of our aerodromes. The final report was presented to the NCAA and Avinor in October 2001. The AEA report stated that the level of safety at Norwegian aerodromes was significantly lower than the target level of safety (TLS) accepted by the ICAO and adopted by the Norwegian Department of Transport and Communications.

The risk analysis concluded with these main points:

- Precision approaches are safer than visual approaches (AEA recommended larger strips/RESA for visual approaches).
- Ground safety margins (strip dimensions) are independent of runway lengths (i.e. code number, runway requirement for different aircraft).
- Requirements are to be dictated by aircraft wingspan and wheelbase (aircraft code letter).

These conclusions are in contrast to the assumptions in Annex 14 stating:

- Visual approaches are safer than instrument approaches (more stringent strip requirements for instrument runways).
- Ground safety margins (strip dimensions) are dependent on code number, i.e. runway lengths required by different aircraft.
- Few of the requirements are dictated by aircraft wing span/wheelbase (aircraft code letter).

In 2002, with acceptance from the Department of Transport and Communications, the NCAA started applying the requirements concluded in the AEA report. From this point on, almost every part of ICAO Annex 14 standards and recommendations were made standard requirements in Norwegian national regulations. Since it was impossible in a short time to implement such a comprehensive change in the regulations, the NCAA made a risk assessment of every airport to find actions that could be implemented in a short time to enhance the safety level (reduce the risk). The result was a series of special requirements for operators performing commercial transportation at our airports. These were implemented in 2002. The following are examples of requirements that are valid for several airports in Norway:

- The operator shall stipulate special crew qualification requirements (Cat. C, confer JAR OPS 1.975).
- The operator shall stipulate special limitations with regard to upper wind.
- The operator shall stipulate special limitations with regard to surface wind.
- The operator shall stipulate special requirements with regard to runway status.
- Operators without steep approach approval shall describe how the flight crew should use the visual aids (visual approach guidance system and runway markings).
- Departure procedures, take-off minima and take-off weight limitations should be documented.
- There will be a requirement to document fulfilment of the special requirements to the Norwegian CAA prior to commencing operations.

The major airlines were of the opinion that Norway should not implement requirements more stringent than the ICAO standards. Most of the new requirements were agreed upon and accepted by Avinor. However, Avinor could not find the requirement of 70m width of the graded part of the safety area for runways with code number 1 and 2, and the transitions beyond the graded part of the strip and RESA justified through the analysis that were done. In Avinor's opinion, these changes were of such a magnitude that an assessment of the consequences according to the law of "building constructions" should have been done before implementation.

The width of the graded part of the safety area was, after additional risk analyses, altered by the NCAA to 40m for runways with code numbers 1 and 2 and 75 m for code numbers 3 and 4. The opposed requirements for transverse and longitudinal slopes beyond the graded part of the strip and RESA were implemented in the national requirements. These requirements are not compliant with Annex 14 recommendations. Figures 10.4, 10.6, 10.10 and 10.12 give some details.

The Current Norwegian Regulation (BSL E) Compared to Annex 14 Standards and Recommendations

In addition to the above mentioned, there are the following differences from Annex 14.

Runway Strip and RESA

The current element dividing runways into instrument and non-instrument in Annex 14 is removed from the Norwegian regulations:

- Runway strips for all runways – both instrument and non-instrument – are the same in the Norwegian regulation.
- Runway code number 1 and 2: 75m to each side of runway centre line, whereby 40m graded.
- Runway code number 3 and 4: 150m to each side of runway centre line, whereby 75m graded.
- The notion "Runway End Safety Area – RESA" is removed from the Norwegian regulations and the RESA is part of the (extended) strip.

All non-precision approach runways shall have a 300m strip in front of threshold for runway codes 3 and 4 and 180m for runway codes 1 and 2. All precision approach runways shall have at least a 60m strip in front of threshold. All runways shall have a 300m strip beyond the runway end for runway codes 3 and 4 and 180m for runway codes 1 and 2. The width of (new) runway strip before the threshold and beyond the runway end shall be the same as the width of runway strip along the runway (unlike today's RESAs in Annex 14).

The reason for the requirement "the runway strip before the threshold and beyond the runway end shall be the same as the width of runway strip along the runway" is understood to be the recommendation in Annex 14 where it is said that "The width of a runway end safety area should, wherever practicable, be equal to that of the graded portion of the associated runway strip."[1]

Precision Approach-like Runway

"Precision approach-like runway" was introduced in the new regulation based on the findings from the AEA analyses concluding that precision approaches are safer than visual approaches. The NCAA has accepted that runways with at least "visual approach slopes indicator", "marked aiming point" and/or other equipment assuring that the touchdown point is of similar accuracy as for category 1 operations can be certified based on the same regulation as for precision approach runways.

The most important effect of this is the possibility to reduce the safety area before threshold to 60m. For a non-precision runway the length would have been 180m for code number 1 and 2 runways and 300m for code number 3 and 4 runways (ref. states responsibility under applicability in the foreword).

---

1 The non graded part of the (new) strip before threshold and beyond runway end in the Norwegian regulation is not a recommendation in Annex 14. The control of obstacles in this area for the purpose of safe overflying is also taken care of by the obstacle limitation surfaces.

The possibility in the requirement to certify runways as "precision approach-like runways" with a 60m safety area before threshold has made it possible during re-certification to maintain the declared distances at almost every aerodrome. At many aerodromes with code numbers 1 and 2 we have even increased the declared distances. All runways at Avinor's airports are now certified as precision approach runways or precision approach-like runways.

All in all there are few differences between the new Norwegian regulations and the standards and recommendations in Annex 14. However the differences in the requirements on the transverse and longitudinal slopes beyond the graded part of the strip and RESAs are costly and difficult to comply with. It is also difficult to assess the risk reduction implied in these regulations compared to Annex 14 recommendations since the longitudinal and transverse slopes are steeper than those in Annex 14. However the Norwegian regulations also require maximum slopes of 20 per cent outside the graded part. Furthermore, the construction of these transverse and longitudinal slopes is supposed to prevent fatal accidents of aircraft overrunning the graded part of the strip or RESA. The requirement is difficult to comply with since there is no guidance in the regulations on how to comply with these requirements. The risk reduction is also considered difficult to justify. After several discussions with the NCAA, Avinor suggested alternative ways of complying with the regulation and also how to limit the requirement of "no fatal accident of aircraft overrunning the graded part of the strip and RESA".

The alternative ways of complying with the regulation are mainly based upon Annex 14 recommendations and the results from the risk analysis made by AEA Technology plc and Det Norske Veritas. These alternative ways of complying are considered to meet the TLS and are accepted by the NCAA without the need for (further) mitigation. Figures showing the Norwegian regulation and the alternative ways of complying are shown in Figures 10.5, 10.7, 10.11 and 10.15.

The Major Challenges at Avinor's 46 Airports

As mentioned in the foreword, the problems of complying with Annex 14 are usually related to the physical dimensions of the runway and the safety areas around the runway, and the needs for obstacle-free zones for safe approaches, landings and take-offs. We will now focus on Avinor's approach for implementing the regulations related to these topics:

Safety Areas

Figure 10.1 shows the situation according to the previous regulation – before re-certification. Most of the short-field aerodromes were published as 1C non-instrument runways even though the declared distances were > 800m. This was due to the cost implied in meeting the strip requirements for 2C (40m). There was no requirement as to the slopes outside the strip.

Figure 10.2 shows the typical layout at Norway's larger aerodromes. Some runways with code numbers 3 and 4 have 240m RESA. The requirement for longitudinal and transverse slopes should follow the recommendations in Annex 14 (see Figures 10.11 and 10.13).

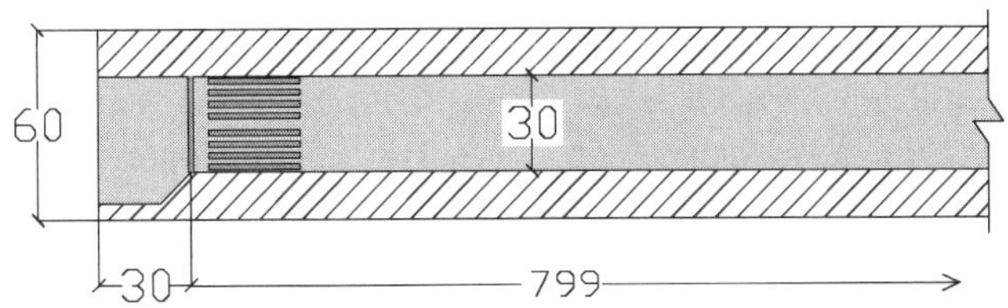

Figure 10.1 Code number 1: non-instrument runway with strip, no RESA – typical layout at Norway's short-field aerodromes[2]

Note: Most of the short field aerodromes were published as 1C non instrument runways even though the declared distances were > 800m. This was due to the cost implied in meeting the strip requirements for 2C (40m). There was no requirement to the slopes outside the strip.

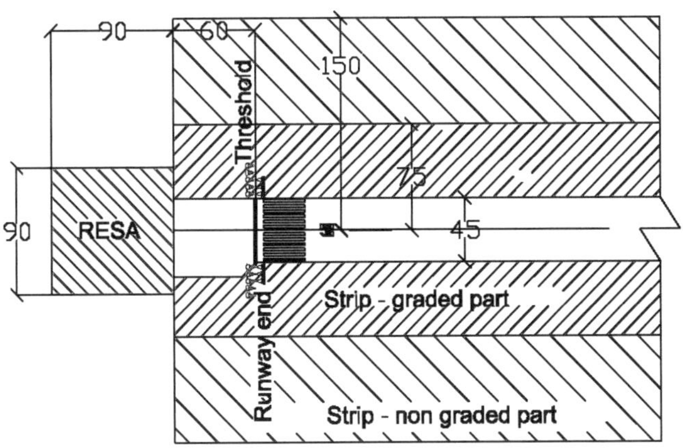

Figure 10.2 Code numbers 3 and 4: instrument runway with strip/RESA – typical layout at Norway's larger aerodromes

Note: Some runways with code number 3 and 4 have 240m RESA. The requirement for longitudinal and transverse slopes should follow the recommendations in Annex 14. See figure 10.11 and 10.13.

2 Avinor is the source of all figures and tables in this chapter unless stated otherwise.

The situation to be reached according to the new regulation – runways with codes 1 and 2 Figure 10.3 shows new physical characteristics of code number 1 and 2 precision approach or precision approach-like runways with strip. The strip also includes the RESA in the new regulations.

Transitions between differing slopes for code 1 and 2 runways Figure 10.4 shows the longitudinal slopes according to BSL E 3-2. The slopes outside 180m should be graded at a length similar to the nose/wheelbase of the largest aircraft using the runway. When the graded part of the safety area is partly water the length of the graded part should extend to 1m below mid-water level. Figure 10.5 shows the longitudinal slopes according to Annex 14, which are accepted as an alternative layout without further mitigation.

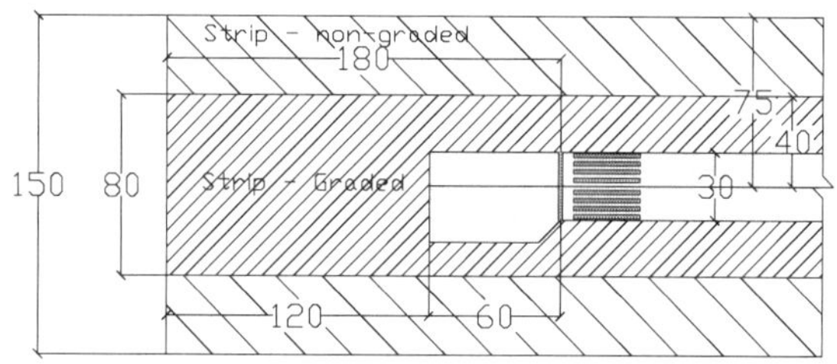

Figure 10.3 Code number 1 and 2 precision approach or precision approach-like runways with strip – new physical characteristics

Note: The strip is also including the RESA in the new regulations.

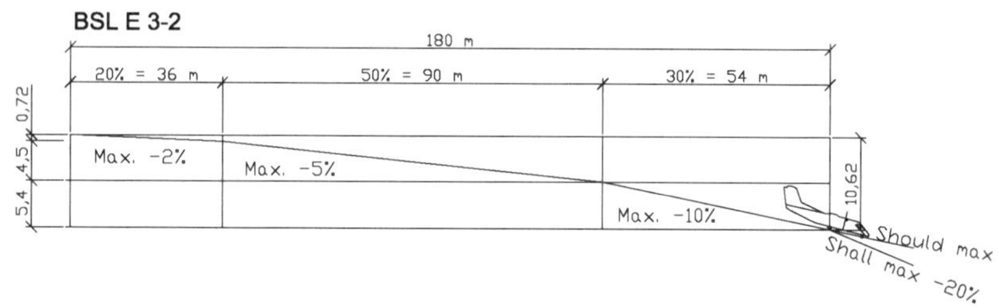

Figure 10.4 Longitudinal slopes according to BSL E 3-2

Note: The slopes outside 180 m should be graded at a length similar to the nose/wheelbase of the largest aircraft using the runway. When the graded part of the safety area is partly water the length of the graded part should extend to 1 m below mid water level.

Figure 10.6 shows the transverse slopes according to BSL E 3-2. Here the slopes outside 40m of the graded part of the strip should also be graded at a length similar to the nose/wheelbase of the largest aircraft using the runway. Figure 10.7 shows the modified transverse slopes based on Annex 14. This is an alternative layout accepted without further mitigation. The layout is based on Dash 8-100 code C aircraft as the largest aircraft using the runway and there are no requirements as to the slopes outside 55m. Figure 10.8 shows the transverse slopes according to Annex 14 but not used in Norway.

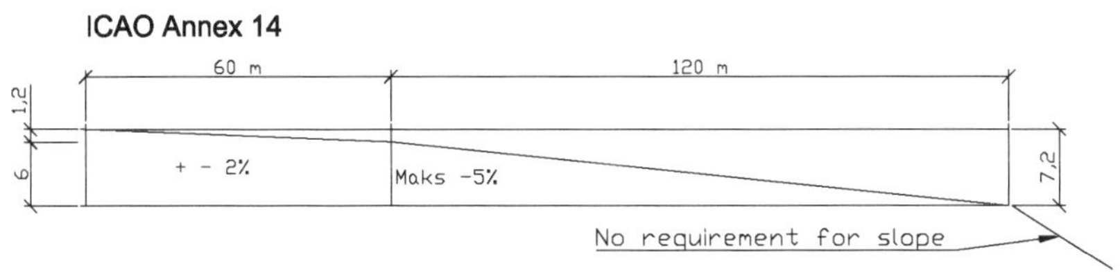

Figure 10.5 Longitudinal slopes according to Annex 14

Note: The slopes in Annex 14 are accepted as an alternative layout without further mitigation.

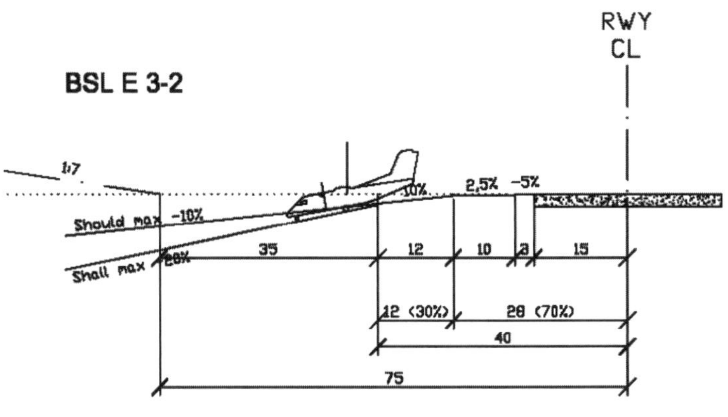

Figure 10.6 Transverse slopes according to BSL E 3-2

Note: The slopes outside 40m of the graded part of the strip should also be graded at a length similar to the nose/wheelbase of the largest aircraft using the runway.

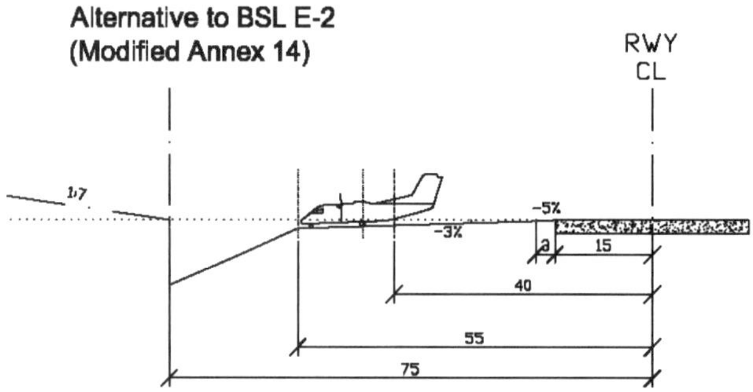

Figure 10.7 Modified transverse slopes based on Annex 14

Note: This is an alternative layout accepted without further mitigation. The layout is based upon Dash 8-100 code C aircraft as the largest aircraft using the runway. No requirement to the slopes outside 55m.

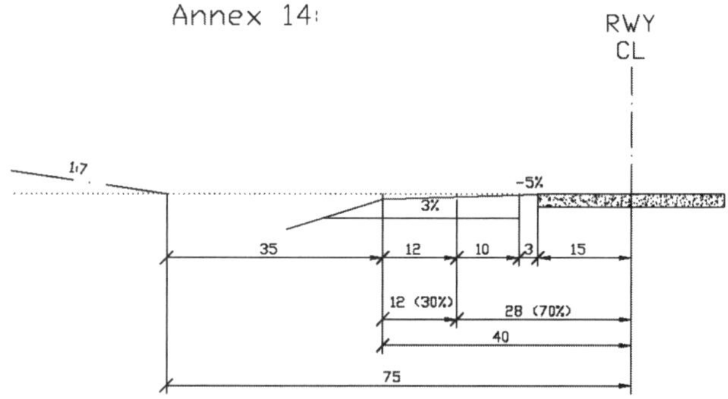

Figure 10.8 Transverse slopes according to Annex 14 – not used in Norway

Note: Layout in accordance with Annex 14 for comparison only; not used in Norway for code 1 and 2 runways.

Transitions between differing slopes in the new regulation for code 3 and 4 runways Figure 10.9 shows code numbers 3 and 4, precision approach or precision approach-like runways with strip – new physical characteristics. The strip also includes the RESA in the new regulation. Figure 10.10 shows the longitudinal slopes according to BSL E 3-2. Here the slopes outside 300m should be graded at a length similar to the nose/wheelbase of the largest aircraft using the runway. When the graded part of the safety area is partly water the length of the graded part should extend to 1m below mid-water level. Figure 10.11 shows the longitudinal slopes according to Annex 14, which accepts slopes as an alternative layout without further mitigation.

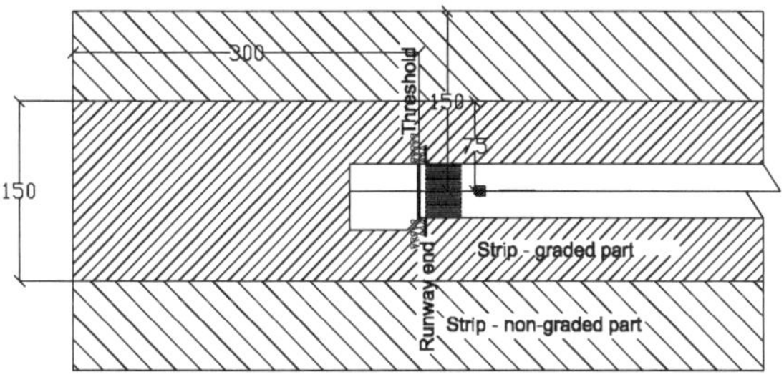

Figure 10.9 Code numbers 3 and 4: precision approach or precision approach-like runways with strip – new physical characteristics

Note: The strip is also including the RESA in the new regulation.

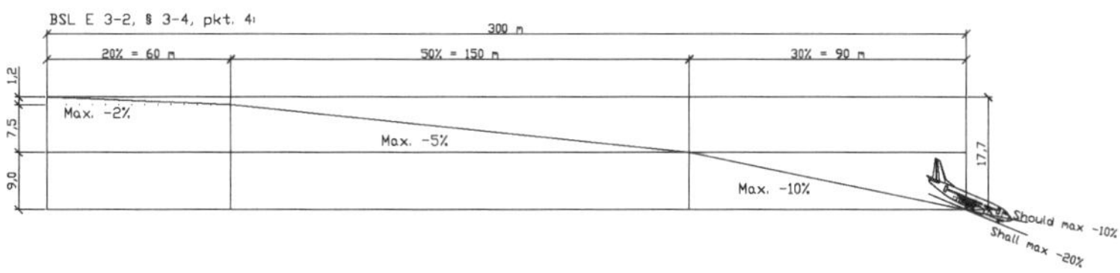

Figure 10.10 Longitudinal slopes according to BSL E 3-2

Note: The slopes outside 300 m should be graded at a length similar to the nose/ wheelbase of the largest aircraft using the runway. When the graded part of the safety area is partly water the length of the graded part should extend to 1 m below mid water level.

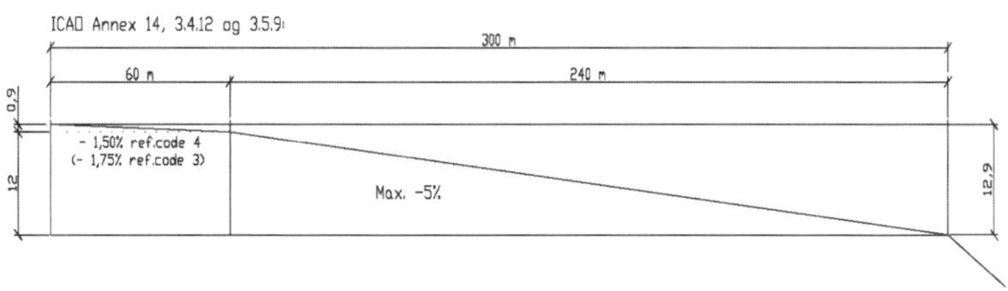

Figure 10.11 Longitudinal slopes according to Annex 14

Note: The slopes in Annex 14 are accepted as an alternative layout without further mitigation.

Figure 10.12 shows the transverse slopes according to BSL E 3-2. Here the slopes outside 75m of the graded part of the strip should also be graded at a length similar to the nose/wheelbase of the largest aircraft using the runway. Figure 10.13 shows the transverse slopes according to Annex 14. This is an alternative layout accepted without further mitigation, and is similar to the recommendation in Annex 14. This is accepted for runways with code numbers 3 and 4 only.

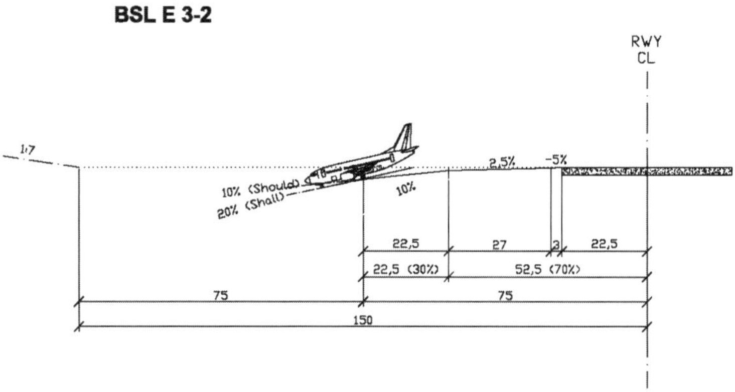

Figure 10.12 Transverse slopes according to BSL E 3-2

Note: The slopes outside 75m of the graded part of the strip should also be graded at a length similar to the nose/wheelbase of the largest aircraft using the runway.

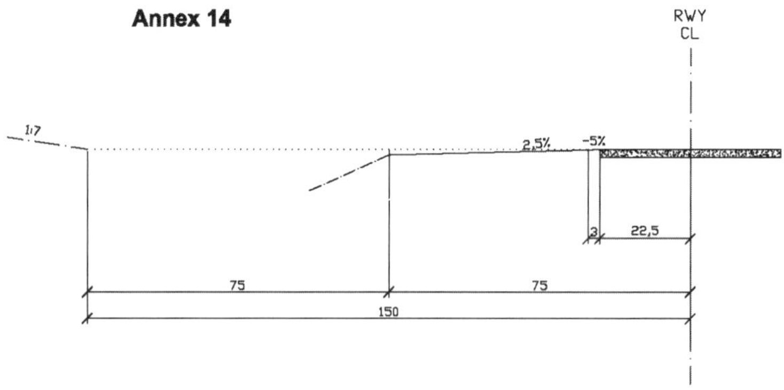

Figure 10.13 Transverse slopes according to Annex 14

Note: This is an alternative layout accepted without further mitigation, and is similar to the recommendation in Annex 14. This is accepted for runways with code number 3 and 4 only.

Obstacle Limitation Surfaces (OLS)

Changes in the OLS Since all the runways with code numbers 1 and 2, and also some of the runways with code numbers 3 and 4, are changing from non-instrument to precision approach/precision approach-like category 1 runways, it causes a considerable increase in the dimensions of the OLS and a severe deterioration with regard to obstacles penetrating the surfaces.

Obstacle restriction and removal Annex 14 states that existing objects above any of the surfaces required should as far as practicable be removed except when, in the opinion of the appropriate authority, the object is shielded by an existing immovable object, or after aeronautical study it is decided that the object would not adversely affect the safety or significantly affect the regularity of operations of aeroplanes.

The Norwegian regulation (BSL E 3-2) requires identification, assessment, mitigation and marking of obstacles when:

- Obstacles are defined according to different obstacles limitation surfaces.
- Terrain and objects that penetrate the limitation surfaces shall be considered as obstacles and be assessed according to the regulation in this chapter.
- All fixed and moveable objects that cannot be removed shall be assessed and, if found necessary, mitigation implemented.
- Target level of safety is to be maintained when mitigation is implemented.

How to Achieve Certification Compliance with the New Regulation

Conclusions from an assessment of the situation before certification can be stated as follows:

- All Norwegian aerodromes face terrain/obstacles in conflict with the OLS.
- Runway safety areas are limited by topography.
- It is often impossible or impracticable to comply fully with the regulations.
- Regulations do open up alternate means of compliance (mitigation), but the target level of safety has to be maintained.
- It is difficult to maintain present regularity and punctuality when mitigation is implemented.
- Many of the special requirements for operators performing commercial transportation should be continued.

Avinor's objectives and actions to meet the regulations are listed below.

Objectives

- Risk assessment of obstacles and safety areas of all 46 Avinor airports to be implemented to find effective actions for risk reduction (mitigation) and cost-effective ways to reduce risk.
- Compliance versus mitigation to be considered based on cost/benefit analysis, avoiding mitigation that reduces regularity and punctuality of scheduled passenger traffic.
- Close cooperation with the airline operators and the regulator.

Actions

- development of a model for risk assessment of obstacles and safety areas;
- development of a model for considering mitigation;
- consult operators when mitigation is introduced.

The Risk Assessment Model – the Approach and Departure Risk Model (ADRM)

A risk calculation model for approach and departure at airports was developed – the Approach and Departure Risk Model (ADRM) – by Det Norske Veritas (DNV) and Avinor in close cooperation with the major airlines (aircraft in brackets):

- SAS (54 B-737 and 6 F50)
- Norwegian (24 B-737)
- Widerøe (30 Dash-8 Q-100/300/400)
- Lufttransport (14 Beech 200/Do 228).

The model assesses the risk in different scenarios and phases of the flight in relation to hazards that can lead to/cause hull loss. One scenario and phases of a flight can be illustrated as an ILS approach with circling and landing.

- Approach/landing – 4 phases:

    1. instrument approach from minimum safety altitude (MSA) to minima
    2. visual to approx. 100ft (depending on aircraft type)
    3. 100ft to touchdown
    4. roll-out to full stop.

- Take-off/departure – 3 phases:

    1. ground roll
    2. climb from lift-off to 400ft above aerodrome (AD)
    3. climb from 400ft to MSA.

The risk assessments are related to the following features in the model:

- obstacles, like terrain, objects on airport;
- safety areas concerning strip, RESA;
- differences between aircraft, like wingspan, equipment and performance;
- navigation aids; characteristics like precision, non-precision, visual;
- visual aids like approach and runway lighting, markings;
- RWY condition characterised by friction, cross-wind;
- local weather characteristics like wind, visibility, wind shear/turbulence;

- ATC interaction like surveillance capabilities;
- consequences like hull loss, number of passengers, probability of survival.

Figure 10.14 gives a risk and event overview, while Table 10.1 gives an example of an account for one scenario and has the following information and characteristics:

- It visualises risk distribution per phase and per unwanted event.
- It helps to identify the large main risk contributors.
- The parameters can be easily changed.
- Recalculation can be made instantly, allowing for sensitivity check and identifying the potential in various risk-reducing measures.

The results from Table 10.1 can be used:

- for drawing an overall risk picture to prioritise resources;
- by airport operators to address the importance of:
  - navigation aids
  - lighting system
  - safety areas and
  - risk-reducing measures (discuss with regulator).
- by airlines to assess the importance of:
  - weather restrictions
  - mandatory training
  - choice of default approach.
- authorities/regulators can use the information as a requirement for approval.

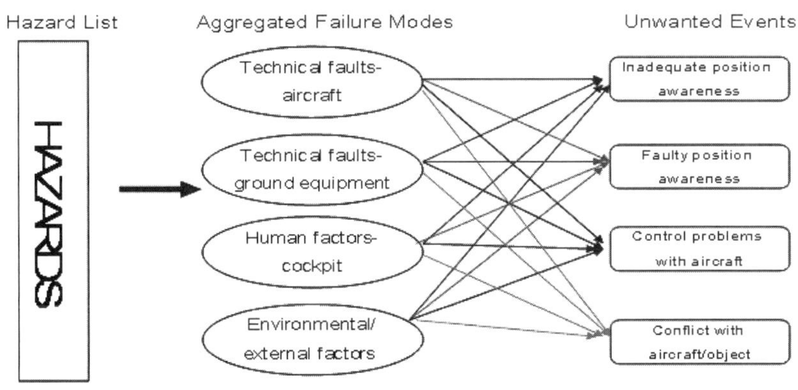

Figure 10.14 Risk elements and events – an overview

Table 10.1 Risk account for one scenario – an example

Top event Unwanted event	Phase	Top event (per landing)	Hull loss frequency (per landing)	Distribution of total hull loss frequency
(i) Inadequate position awareness	1	1,87E-04	1,06E-09	**0,6 %**
	2	8,79E-04	8,79E-08	**48,1 %**
	3	3,03E-05	2,95E-09	**1,6 %**
	4	2,60E-05	3,18E-09	**1,7 %**
	Sum 1-4	-	**9,51E-08**	**52,0 %**
(ii) Faulty position awareness	1	6,06E-05	3,93E-11	**0,0 %**
	2	4,90E-05	1,81E-10	**0,1 %**
	3	5,25E-07	5,25E-10	**0,3 %**
	4	2,80E-06	3,59E-10	**0,2 %**
	Sum 1-4	-	**1,10E-09**	**0,6 %**
(iii) Control problems - aircraft	1	9,75E-07	9,08E-12	**0,0 %**
	2	4,90E-04	6,27E-08	**34,3 %**
	3	6,01E-05	8,98E-09	**4,9 %**
	4	6,59E-05	1,48E-08	**8,1 %**
	Sum 1-4	-	**8,65E-08**	**47,4 %**
Sum top events (i) - (iii)	1	-	**1,11E-09**	**0,6 %**
	2	-	**1,51E-07**	**82,5 %**
	3	-	**1,25E-08**	**6,8 %**
	4	-	**1,84E-08**	**10,1 %**
	Sum 1-4	-	**1,83E-07**	**100,0 %**

Mitigation – Assessment of Deviation from the Regulations

Although the ADRM model for risk assessment of obstacles and safety areas was used for assessing different types of mitigation, it was found too time-consuming to use for minor deviations from the requirements on the physical characteristics of the runway, taxiway and safety area, and on deviations from the requirements on aids for navigation, markings and lights (e.g. their positioning and frangibility). Therefore a simple model for assessing these deviations based on weighting different actions was established by Avinor. This model is also accepted and used by the NCAA to quantify deviation and mitigation.

However the findings and conclusions from the ADRM risk assessments of Avinor's 46 aerodromes shows that the most effective action/mitigation to reduce risk is to establish:

- straight-in precision approach with directional and vertical instrument guidance (like ILS, Scat 1);
- enhanced visual aids during circling and landing;
- enhanced wind and turbulence warnings;
- improved maintenance of runways during winter operations.

For risk reduction we also found that proactive actions related to flight operations were considerably more efficient and cost effective than actions connected to improvement of the safety areas. These findings/conclusions made us start research and development programmes and tests on:

- lighted aiming point and lighted runway centre line;
- balked landing guidance lights on short runways;
- enhanced wind and turbulence warnings.

Lighting of Runway Centre Line and Aiming Points

Inset lights are installed (or are being installed) on the runway center line and the aiming points at most Norwegian airports. The intention is to improve flight safety by making the markings visible under all conditions, including darkness and when covered by snow or ice. The lights are located along the centre line with the same spacing as and perpendicular to the runway edge lights. On the aiming points there are six lights on each marking, three lights on each of the two stripes, equally spaced along the centre line of the stripe.

The lights on the runway centre line are fixed lights showing variable white from the threshold to the runway end. The lights on the aiming point are fixed lights showing variable white in the direction of approach to the runway. The intensity and beam spread of the lights on the runway centre line and the aiming points are in accordance with the specification for runway centre line lights in ICAO Annex 14, Volume I. The intensity of lights on the centre line and aiming points is adjusted together with the runway edge lights.

Information about where the lights are installed is given for each airport in Aeronautical Information Publication (AIP) Norway, Section AD2, Table 2.14.

Balked Landing Guidance Lights

The balked landing guidance lights were developed based on a recommendation from the safety investigation board after several veer-offs at our short field aerodromes. Balked landing guidance lights are installed at some Norwegian airports. The intention is to give the pilot information about the aircraft's position relative to touchdown.

Balked landing guidance lights are fixed unidirectional lights showing yellow in the direction of approach to the runway. The intensity and beam spread are in accordance with the specification for yellow runway edge lights in ICAO Annex 14, Volume I. The intensity of the balked landing guidance lights is adjusted together with the runway edge lights.

Balked landing guidance lights are symmetrically disposed about the runway centre line at a specified distance beyond the runway threshold in two groups, i.e. wing bars. Each wing bar has two lights. The innermost light is placed 4m outside the runway edge lights and the gap between the two lights is 4m. Information about where the lights are installed and the distance beyond the runway threshold is given for each airport in AIP Norway, Section AD2, Table 2.14. Figure 10.15 shows the lighting systems implemented at all Norway's runways with code numbers 1 and 2.

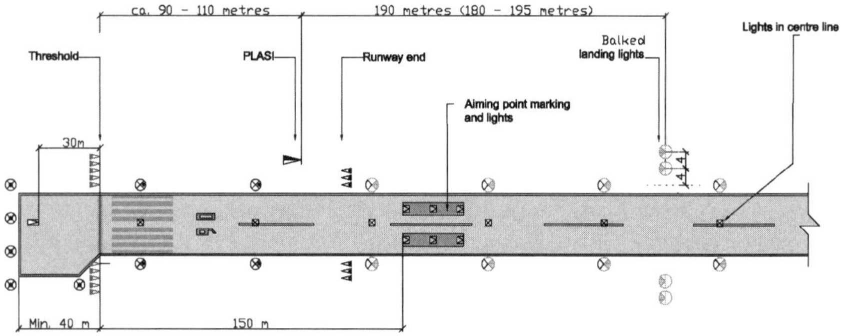

Figure 10.15 Lighting systems, code 1 and 2 airports

Note: For runways with code numbers 3 and 4, the same lighting system applies except when requirements in BSL E 3-2 have to be met for low-visibility operations or Cat II/Cat III operations.

Turbulence Warnings

Figure 10.16 shows turbulence forecast at Sandane Airport, both map and cross-section. Meteorological data from the weather model HIRLAM are processed thorough a terrain data model developed by the research organisation SINTEF and the Norwegian Meteorological Institute (NMI), and the results are presented to the pilots through the Internet Planning Pilot Centre (IPPC). The system is approved by the NCAA for operational tests at seven airports and is expected to be implemented at 30 aerodromes in the near future. The system is highly recommended by operators and is one of the actions accepted by the NCAA for mitigation at airports with obstacles/terrain penetrating the OLS and for deviations on the safety areas.

Mitigating Actions

At every Norwegian airport except Oslo (OSL), there is a combination of mitigating actions to compensate for deviations from the regulation. Due to the high risk reduction of some actions/means, some are made compulsory at Avinor's aerodromes. Table 10.2 shows a list of selected mitigations.

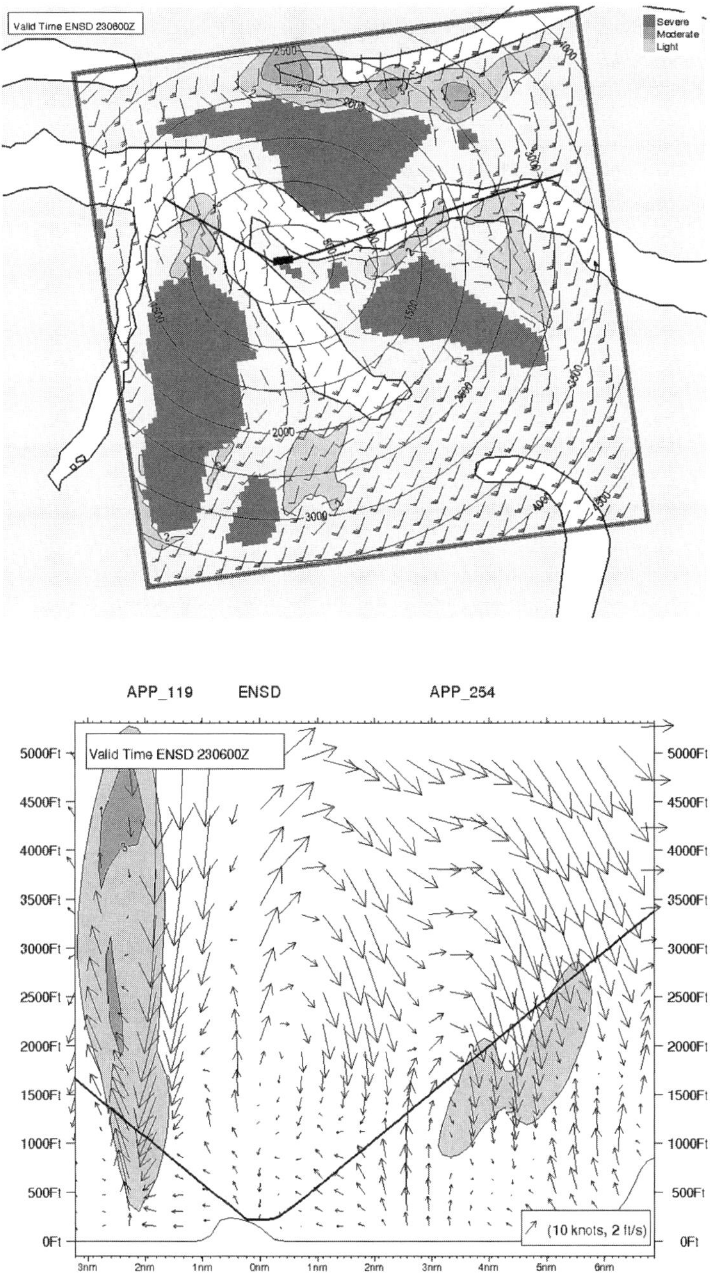

Figure 10.16 Turbulence forecast at Sandane Airport, map and cross-section

Note: Forecast for 23 October 2008. Grey circles: height above sea level in ft.

Table 10.2 Selected risk-mitigating actions

Lighted aiming point and lighted runway centre line (compulsory)
Balked landing lights (compulsory on short runways)
Runway condition – different actions (i.e. minimum braking action, no contamination)
Special requirements for operators
Enhanced wind and turbulence warnings
ILS or SCAT-1
Safety areas better than the requirements (e.g. for parts of the runway)
Engineered Material Arresting Systems (EMAS)
Enhanced visual references (e.g. leading lights, circling lights, obstacle lights, enhanced approached lightning)
Uninterrupted power supply, e.g. for circling and approach lights, obstacle lights
Closing of circling procedures
Grooving of runway (compulsory)
Special procedures (ATC and/or others)
Obstacle charts Type B (compulsory at airports with special requirements for operators)

Current Status

All 46 aerodromes are now certified according to the new regulation. Upgrading (construction work) at the aerodromes is estimated to be finished in 2013. During spring 2009, 16 of the 46 airports were upgraded. The total cost for upgrading all 46 aerodromes is estimated to be NOK 3.5–4 billion (2008). Table 10.3 shows the runway dimensions (landing distance) and passenger movements in the year 2007.

Avinor is the operator of all these aerodromes, and all the cost is covered by Avinor through cross-subsidisation. See Chapter 11 for a further description of Avinor's activities and airport locations.

Table 10.3 Airports and runway (rwy) landing distances

Airport	Landing distance available in metres after upgrading (*under revision)	Airport	Landing distance available in metres after upgrading (*under revision)
Large airports			
Stavanger	*Rwy 11/29 – 2199m *Rwy 18/2496m *Rwy 36/2556m	Bodø	Rwy 07/25 – 2794m
Bergen	Rwy 17/2550m Rwy 35/2525m	Tromsø	Rwy 01/2392m Rwy 19/2004m
Trondheim	Rwy 07/2446m Rwy 27/2424m		
Medium-sized airports			
Kristiansand	Rwy 04/1920m Rwy 22/1870m	Bardufoss	Rwy 10/1970m Rwy 28/2443m
Haugesund	Rwy 14/32 – 2000m	Alta	*Rwy 11/1851m *Rwy29/1907m
Åleseund	Rwy 07/2014m Rwy 25/1978m	Lakselv	Rwy 17/2696m Rwy 35/2604m
Molde	Rwy 07/25 – 1980m	Kirkenes	Rwy 06/1695m Rwy 24/1605m
Kristiansund	*Rwy 07/1780m, *Rwy 25/1760m	Svalbard	Rwy10/2138m Rwy 28/2138m
Harstad/Narvik	Rwy 17/2716m Rwy 35/2750m		
Regional airports			
Fagernes	Rwy 15/1843m Rwy 33/1836m	Leknes	Rwy 16/34 – 829m
Sogndal	Rwy 06/25 – 870m	Svolvær	Rwy 01/807m Rwy 19/777m
Florø	Rwy 07/25 – 1199m	Stokmarknes	Rwy 09/829m Rwy 27/799m
Førde	Rwy 07/25 – 890m	Narvik	Rwy 01/830 m Rwy 19/824m
Sandane	Rwy 09/27 – 760m	Andøya	Rwy 14/32 – 2468 m
Ørsta-Volda	Rwy 06/24 – 890m	Sørkjosen	Rwy 15/33 – 799m
Røros	Rwy 14/32 – 1580m	Hasvik	Rwy 01/819m Rwy 29/799m
Namsos	Rwy 07/25 – 800m	Hammerfest	Rwy 05/850m Rwy 23/824m
Rørvik	Rwy 04/22 – 830m	Honningsvåg	Rwy 08/26 – 800m

Table 10.3 ***Concluded***

Airport	Landing distance available in metres after upgrading (*under revision)	Airport	Landing distance available in metres after upgrading (*under revision)
Brønnøysund	Rwy 04/22 – 1199m	Mehamn	Rwy 17/830m Rwy 35/850m
Sandnessjøen	Rwy 03/21 – 881m	Berlevåg	Rwy 06/24 – 829m
Mo i Rana	Rwy 14/32 – 830m	Båtsfjord	Rwy 03/21 – 810m
Mosjøen	Rwy 16/34 – 829m	Vadsø	Rwy 08/26 – 830m
Røst	Rwy 03/21 – 830m	Vardø	Rwy 15/33 – 905m

References

AEA Technology plc (2001). Final Report on the Risk Analysis in Support of Aerodrome Design Rules. Report No. AEAT/RAIR/RD02325/R/002.

Det Norske Veritas AS (2005). Safety Areas Along Runways. Report No. 2005–1175.

ICAO (2004). Annex 14 to the Convention on International Civil Aviation. Fourth edition, July.

NCAA (2001). Norwegian Regulation BSL E 3-2.

NCCA (regular updates). Aeronatical Publication Norway.

Norwegian Metereological Institute. http://www.met.no.

SINTEF (2007). Runway strip width and lateral veer-off risk at Norwegian runways. Report No. STF50 F06120.

Chapter 11

Developing Airports in a Long-Term Perspective – the Case of Tromsø, Norway

Svein Bråthen and Knut Fuglum

Introduction

Airport development is important to facilitate air transport in remoter areas. Local airports are facing new challenges because of International Civil Aviation Organization (ICAO) Annex 14 regulations, new types of aircraft and updated navigation systems such as SCAT-1. Some of the bigger regional airports act as hubs, linking remoter local airports and trunk domestic and international routes.

This chapter will describe the development of airports in a long-term perspective. Tromsø Airport in northern Norway will be used as an example. A brief description of Avinor as responsible for airports and air traffic management will be given. Then we will describe briefly how the need for long-term development of airport infrastructure can be addressed. Finally, we will show how the need for long-term development of Tromsø Airport's infrastructure has been analysed as a part of its Master Plan (MP). The full report is given in Bråthen and Husdal (2008).

Avinor's Operations

Avinor owns, operates and develops airports and air navigation services in Norway, with a focus on safety and security. The company's primary objectives are to facilitate safe and efficient aviation, organize the commercial activities at the airports and to make capacity and expertise available for airlines.

Avinor was established as a limited liability company on 1 January 2003, and it is fully owned by the Norwegian Government through the Ministry of Transport and Communications. Norway's busiest airport, Oslo Airport Gardermoen, is a fully owned subsidiary. Previously, the activity was organized as a state enterprise – Norwegian Air Traffic and Airport Management (NATAM), accountable to the Ministry of Transport and Communications.

Avinor owns and operates 46 airports: 12 of these are in association with the armed forces; 28 of these 46 airports are local short take-off and landing (STOL) ports, mainly served by operators under Public Service Obligation (PSO) (Figure 11.1).

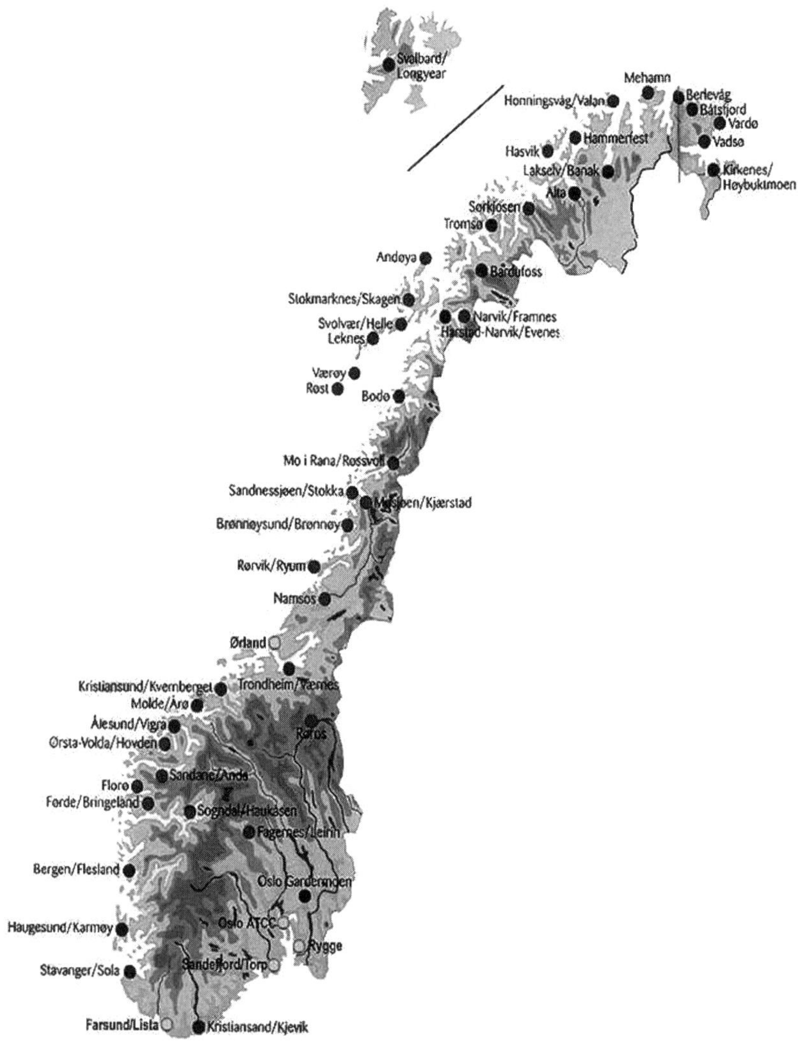

Figure 11.1 Avinor's network of airports

Source: Avinor

The Government subsidizes PSO routes with approximately NOK 500 million per year. For remote and scarcely populated areas, these PSO routes are of great importance. More than 90 per cent of domestic traffic and all international traffic is, however, operated on a purely commercial basis.

Norway is a far stretched and scarcely populated country, with a population of only 4.6 million. The country is traversed by fjords and mountain ranges.

Hence, the topography is one of the reasons why aviation is the most efficient mode of transport on longer distances. One half of the population lives in the Oslo fjord region; 11 per cent of the population lives in northern Norway, where almost 60 per cent of Avinor's airports are located. The airports are well spread out geographically, and they represent a fine-meshed network with good international connections.

Only four airports in Norway with scheduled traffic exceeding 10,000 passengers per year are not owned by Avinor. These are Rygge, Torp and Skien in the south east, and Stord on the west coast. These airports handle some 4 per cent of the total traffic. Torp is by far the busiest, with 1.6 million terminal passengers in 2008.

Avinor is run as an integrated and self-financed airport and air traffic control (ATC) system. Only four airports are significantly financially profitable (Oslo, Stavanger, Bergen and Trondheim). Hence, Avinor relies on heavy cross-subsidies from these busy airports. Lian et al. (2006) have shown that:

- Air travel accounts for 45 per cent of all domestic trips longer than 300 kilometres.
- Norwegians are very frequent users of domestic air transport. The average annual trip rate per capita is 2.3. Most other European countries have less than a third of this trip rate, the only exception being Iceland with around half the level of Norway.
- Almost 70 per cent of all Norwegians have less than 1 hour travelling time by surface mode to an airport with domestic services. Within 2 hours' travelling time, 90 per cent of the population is covered.
- The airport system plays a significant role in industry, trade and welfare. This is described further in Chapter 4.

Figure 11.2 shows the annual number of passengers during the period 1990–2008. The figure shows that after a period with continuous growth until 1999, Norwegian airports experienced a decline in traffic similar to most other countries. From 2003, they recorded a pronounced growth, but with a decline during the last months of 2008 and the beginning of 2009. Key figures for Avinor's activities in 2008 are:

- 42.5 million terminal passengers – 2/3 domestic, 1/3 international, 19.3 million at Oslo Airport, Gardermoen, 1.3 million at the 28 local airports;
- 870,000 aircraft movements.

Figure 11.3 shows how Avinor's revenues have developed in the period 1994–2007. It is apparent that during the last five to six years the importance of commercial activity increased significantly, whereas state aid to local airports decreased.

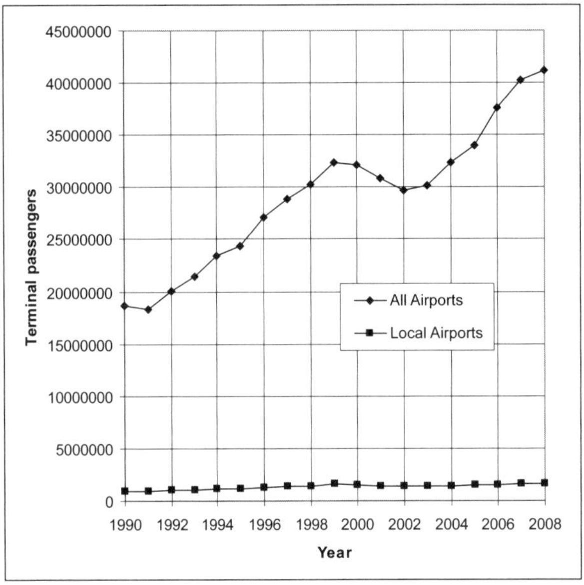

Figure 11.2 Terminal passengers at Norwegian airports 1990–2008

Source: Avinor

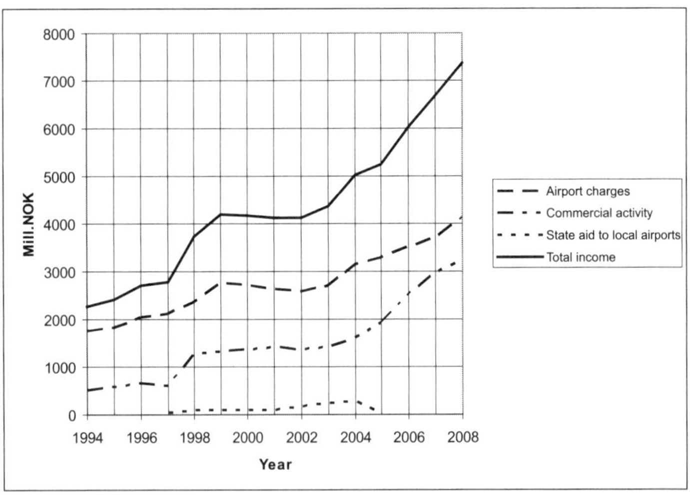

Figure 11.3 Annual revenues 1994–2007 (million NOK)

Source: Avinor

Key figures for Avinor's financial situation in 2007 were:

- operating revenue: NOK 6,700 billion (of which NOK 140 million from local airports) – 55 per cent from airport charges; 45 per cent from commercial activities;
- investments: NOK 1,990 million;
- state aid to local airports: NOK 0;
- dividends to the Government: NOK 397 million;
- requirement for cross-subsidies between airports: NOK 1,073 million;
- 2,850 man-years worked.

Long-term Planning of Airport Infrastructure

Figure 11.2 shows that, despite a few downturns, air traffic has been growing steadily. This raises the need to ensure that the airports are left with enough capacity to serve the expected growth in the number of passengers. On one hand, such long-term planning needs to address the need for expansion and upgrading of terminals, runways, taxiways and other kinds of infrastructure. On the other hand, economic analyses are needed to check whether the planned expansion and upgrading are financially and economically viable. In many cases, such measures are often expensive and all relevant alternatives should be carefully assessed. However, it is a good strategy to ensure the necessary real options in terms of available expansion possibilities to meet a high future traffic growth.

In this chapter the Master Plan for Tromsø Airport, Langnes (TOS) is used as an example of long-term airport planning in remoter areas. The MP contains a technical, financial and socio-economic part, and we will focus mostly on the socio-economic part in this chapter. TOS is the busiest airport in northern Norway, with 1.6 million annual passengers and 40,000 aircraft movements. It serves as a connecting hub for the northern part of the extensive STOL network in Norway, as shown in Figure 11.4. The STOL networks are shown by thicker lines, whereas the main routes are shown by thin lines. A lack of capacity in such local hubs will of course affect large parts of the STOL network as well as some of the main routes.

Tromsø Airport, Langnes (TOS)

Tromsø is located at nearly 70°N in northern Norway. It is the region's largest city, with almost 65,000 inhabitants. The municipality has traditionally been regarded as a fisheries and whaling centre, but in recent decades this focus has changed. Tromsø is now an important centre of culture, health education, research and administration both for the county of Troms and the northern part of Norway. The city also hosts a university.

Figure 11.4 The air transport network in Norway

Source: Adapted from Avinor

The county of Troms has three airports with scheduled traffic: regional airports Tromsø Airport, Langnes (TOS) and Bardufoss (BDU) and local airport Sörkjosen (SOJ). TOS serves around 90 per cent of the traffic to and from the county of Troms. Military activity is important at BDU, but in recent years this airport has also been developed as a gateway to family activities under the name of 'Snowman International Airport'.

There is a fourth airport, Evenes (EVE), on the border with Nordland County. Figure 11.5 shows the different airport locations. Traffic figures in 2008 were:

- Bardufoss: 181,000 terminal passengers, 3,200 aircraft movements;
- Evenes: 514,000 terminal passengers, 9,600 aircraft movements;
- Sörkjosen: 21,000 terminal passengers, 2,000 aircraft movements.

The zones around the airports show the airports' primary catchment areas of one-hour access time by road. There are no significant overlaps in these catchment areas. Hence, these airports are serving mostly different markets.

TOS is located on the western side of Tromsøya, approximately 3.5 kilometres north-west of Tromsø city centre. The airport serves mainly scheduled civil air traffic, but it has also some military activity. TOS is one of Norway's larger regional airports. It serves mainly domestic traffic, but also some international traffic. The runway is 2,400m, which is considered as sufficient for serving relevant aircraft types without any restrictions on weight or number of passengers.

The airport had an annual traffic growth of 6.6 per cent from 1990 to 1999, and it reached about 1.6 million passengers, including transit and transfer, in 1999. Between 1996 and 2003, traffic fell to 1.35 million passengers. However, traffic has grown steadily since then. In 2008, TOS had about 1.6 million terminal passengers (including some 275,000 transfers). Traffic growth is expected to continue, as shown in Figure 11.6.

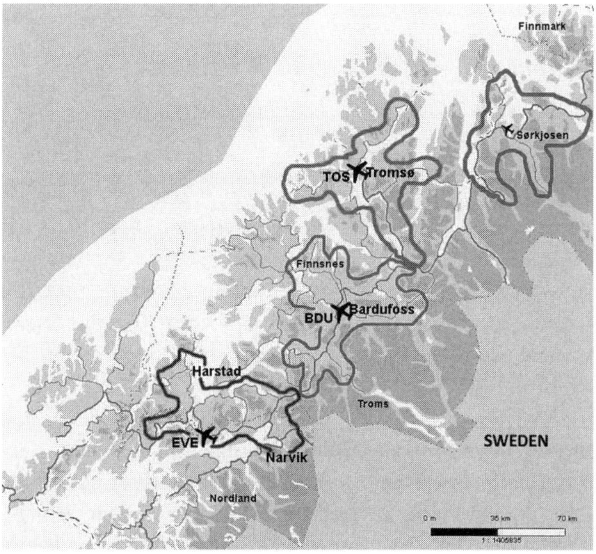

Figure 11.5 Location of Tromsø Airport and adjacent airports

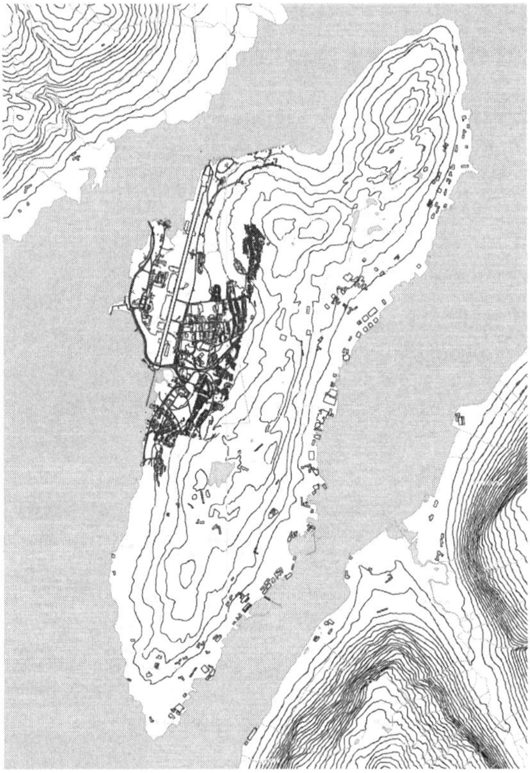

Figure 11.6 Tromsøya with the airport
Source: Avinor

According to these scenarios, accumulated passenger growth will amount to between 15 per cent and 50 per cent between 2007 and 2020, giving 1.8–2.3 million passengers in 2020. The reference (mid) scenario shows an annual growth of 2.3 per cent up to 2020, giving approximately 2 million passengers. Because of an expected increase in load factor and larger aircraft, the growth in aircraft movements is expected to be 1.1 per cent per year in the reference scenario.

Economic Impact Assessment – The Concept

This anticipated growth raises the need for expansion and upgrading of the airport infrastructure. The main and most expensive measures that need to be taken are taxiway extensions, terminal renewals/expansions and aircraft stands, as shown in Figure 11.7. The discounted costs of the Master Plan amount to around NOK 460 million (per 2007). The planning horizon covers the next 40 years.

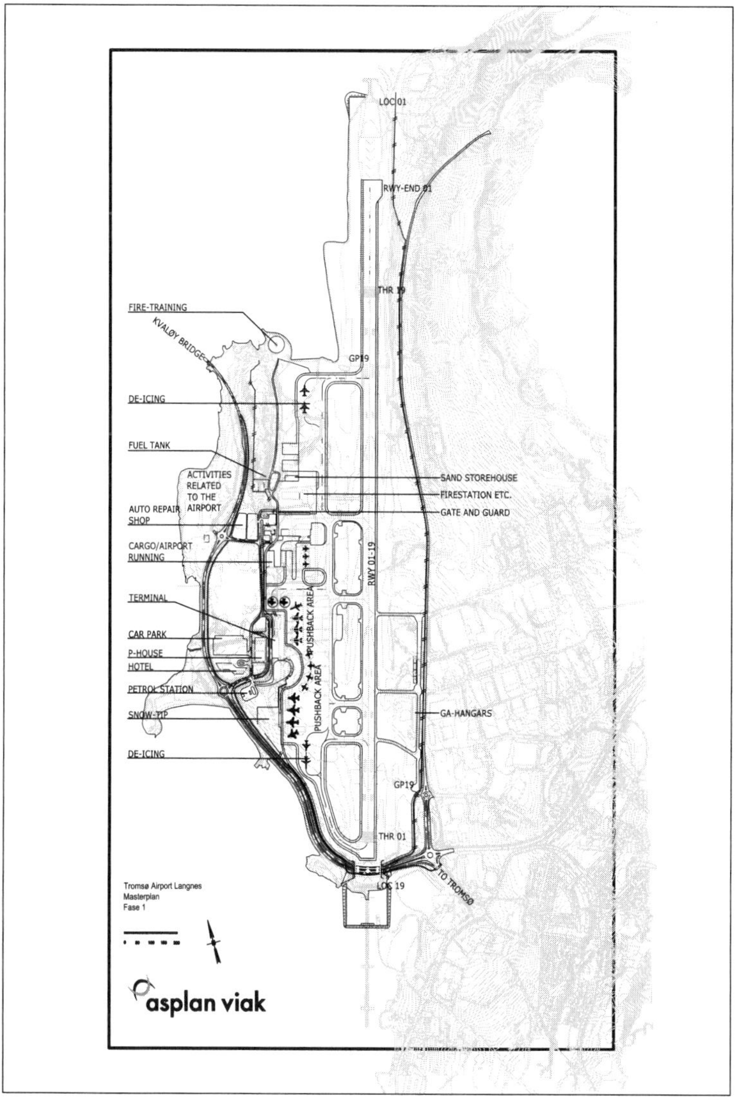

Figure 11.7 The Master Plan for Tromsø Airport

The Master Plan should be subjected to a financial analysis. As mentioned above, Avinor is a limited liability company. Apart from the fact that airside charges are regulated by the Ministry of Transport and Communications, Avinor has full financial responsibility for its own activities. However, financial deficits at the airport level are allowed as long as they can be covered by means of cross-subsidies from financially profitable airports.

In addition to the financial analysis, there is also a need for a socio-economic appraisal, because all the economic effects of air transport are not included in the financial cash flow. In particular, this has to do with the passengers' travel time. If the capacity at TOS is not expanded to serve the future traffic growth, a welfare loss to society is likely to occur because the passengers would need to find either other modes of transport, more distant airports such as BDU and EVE (see Figure 11.5) or refrain from travelling at all. This means that extra travel time is imposed on passengers, without compensation in terms of lower air fares to cancel out the extra travel time costs and other travelling costs.

To understand the effects of scarce future airport capacity if expansion and upgrading of TOS is left undone, a cost benefit analysis (CBA) framework is proposed and used. This framework is used for larger infrastructure investments within Norwegian air transport, and it is described in detail in Bråthen et al. (2000, 2006). The main idea is to calculate the costs of the most reasonable transport alternative to make passengers maintain the same origin-destination pattern as if TOS had been expanded to meet future demand. CBA gives the net changes in the use of real resources such as time, vehicle/vessel operating costs and emissions, measured in monetary terms. The economic effects on passengers economic welfare is measured in terms of changes in consumer surplus (CS), which is the change in difference between the maximum willingness to pay for getting from origin to destination minus the total travel costs of getting there. The total willingness to pay is not affected by airport capacity; however the total travel costs are, if passengers need to find other routes. In short, what we do in a CBA is compare the economic effects of a situation where measures are taken to maintain necessary capacity to allow expected traffic growth – with a situation where the capacity is not expanded at all.

The most important economic effects to be addressed within the context of a Master Plan are:

- Changes in total travel costs if other transport modes have to be used in case the plan is not implemented, often in terms of changes in time costs, costs of longer driving distances to adjacent airports, costs of using, e.g., train services instead etc.
- Environmental effects, like noise and emissions to air.
- Effects on safety. These effects may be particularly important if the road network has to be used instead of air transport. The probability of casualties per passenger kilometre is significantly higher in the road network.
- Investments and operating costs for airlines and airports.

Network effects are relevant when flight corridors and routes are affected. These include effects on the generalized travel costs for passengers in other parts of the network, as well as for different operators. In the case of TOS, these effects are not calculated even if to a certain extent they might strengthen the socio-economic profitability of the Master Plan.

Figure 11.8 shows the change in consumer surplus (CS) if airport capacity is constrained. As mentioned, the change in CS denotes the welfare effects for the passengers, which are the most significant effects in addition to the effects on investment and operating costs. The area below the demand curve denotes the willingness to pay for air transport services. As air fares are reduced, demand increases.

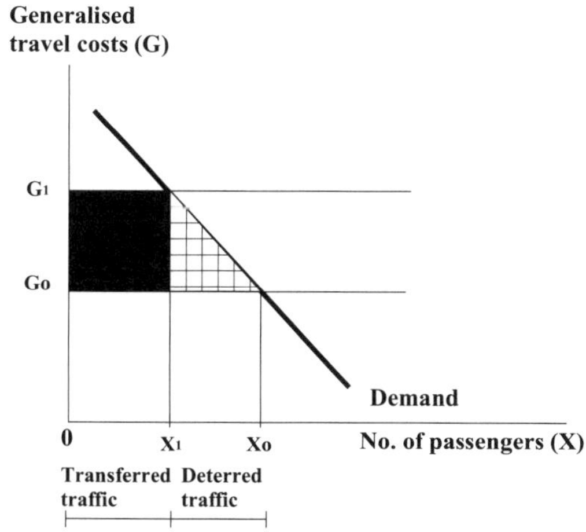

Figure 11.8 Economic loss to air passengers because of inadequate airport capacity

The difference in generalized travel costs causes two effects:

- deterred traffic (shown as $X_0 - X_1$ in the figure because of increased travel costs from constrained airport capacity);
- traffic that will still use an adjacent airport or another transport mode (X_1), but at higher costs.

These two effects are used when calculating the black and crosshatched areas. The black area is the loss in consumer surplus for those who will still travel from an adjacent airport, while the crosshatched area is the loss for those who are deterred because of the higher travel costs. The total value of the net change in CS discounted over 25 years is:

$$N_T = \sum_{i=1}^{25} \frac{X_{1i} + X_{0i}}{2} (G_1 - G_0)(1 + r)^{-i}$$

where:

N_T = net present value of the CS change, 25 years (i=1 … 25);
X_{1i} = traffic that will use an adjacent airport, year i;
X_{0i} = traffic at the airport with capacity constraint, year i;
G_1 = generalized travel costs by going from an adjacent airport;
G_0 = generalized travel costs by going from the constrained airport;
r = discount rate (in Norway 4.5 per cent).

In addition to the changes in passengers CS, we have calculated emission costs and safety effects separately. The sum of all these effects is then compared with the investment and operating costs in the Master Plan to see whether the project is viable from an economic point of view.

Normally, the time span for the analysis is 25 years. Hence, the CBA does not comprise more than the first approximately 2/3 of the Master Plan's horizon. However, because of the discounting procedure, future costs and benefits after 25 years will have relatively little impact on the results. Discounting is done to be able to compare air transport projects with the best alternative use of money on public projects with comparable risk. A standard textbook on economics, such as Varian (2006), explains the reason for discounting in some detail.

We will not go into detail on the various factor values that are used in the CBA, but we will briefly describe some of them. The value of travel time for business travel is set to approximately NOK 350 per hour (2007), while it is around NOK 240 per hour for leisure travel. The value for business travel should in principle reflect the value of using travel time for productive purposes, and hence the value should not be too far from the gross hourly wage rate. These values are applied from Ramjerdi et al. (1997), and the application is shown in more detail in Bråthen et al. (2006). These values will be revised, presumably during 2010. The value for CO_2 emissions is set to NOK 300/tonne, which is higher than e.g. the PointCarbon index, which was around NOK 150 (per July 2008). The value for traffic casualties (deaths) is around NOK 25 million per statistical life (one random life). The annual impact on safety from a given project is often expressed as the *change in probability of casualty* × *the annual number of aircraft movements or vehicle kilometers affected* × *the average number of passengers on board.* Various factor values are discussed in Bråthen et al. (2006).

Economic Impact Assessment of the Master Plan (MP) for TOS

The main point in a socio-economic analysis is to find out whether expansion and upgrading of TOS is economically viable. The MP has three stages, where Stage 1 goes from 2008 to 2016 and Stage 2 from 2017 to 2030. The MP sketches some necessary measures in Stage 3 (2031–2050), but the costs of these measures are not calculated due to the high degree of uncertainty with respect to technological solutions, among other factors. Therefore, the CBA deals with Stages 1 and 2. Omitting Stage 3 does not affect the results in any significant way. As said above, the discounted costs of the measures in Stages 1 and 2 are NOK 460 million. The starting point of the analysis is to estimate the number of passengers who will not be able to use TOS because of capacity constraints if the MP is not implemented. It is worth noting that we have only addressed the situation with the highest traffic scenario in Figure 11.6. The reason is that this scenario determines the necessary array of measures in the MP. If traffic develops below this, the MP will be revised accordingly. There are reasons to stress the importance of paying close attention to the fine-tuning of the relations between traffic level and timing of the investments. This is also done in the MP document. Figure 11.9 shows the expected growth in number of passengers with and without expansion/upgrading.

The passenger numbers are arrivals/departures. The dotted vertical line denotes the life span of the CBA, which is 25 years from 2007 to 2032. The lower line is the estimated traffic if none of the measures in the MP is implemented. The mid line is the estimated traffic if only Stage 1 is implemented. These lines are modelled with an exponentially declining growth rate as the capacity constraint is approached. Without any measures in the MP, the capacity limit is anticipated to be around 1.8 million passengers (lower line). With only Stage 1 implemented, the capacity limit is expected to be around 2.2 million. The upper line is the unconstrained passenger growth (all measures in the MP are implemented). Stage 3 is not considered here, because it will only take effect well beyond the time span of the CBA.

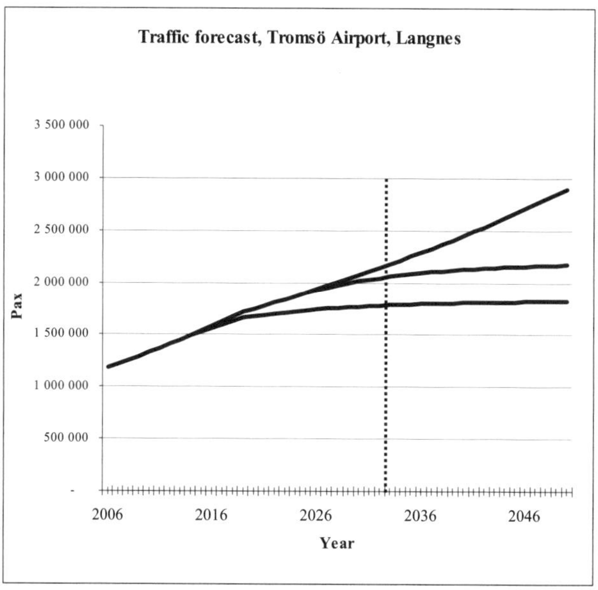

Figure 11.9 Traffic forecast for Tromsø Airport, 2006–2050

The distance between the lower and upper line can be considered as the resulting number of extra passengers that will follow from the implementation of the MP (Stage 1 + Stage 2). The area between the mid line (Stage 2) and the upper line is the extra traffic that can result from the MP Stage 2 alone, of course given that Stage 1 is already implemented. The analysis presented here considers the effects of both Stage 1 and Stage 2 (e.g. approximately 200,000 passengers in 2025). These passengers are now considered as either 'transferred to cheapest alternative transport mode' (X_1 in Figure 11.8) or 'deterred from travelling because of higher travel costs on the cheapest alternative mode' ($X_0 - X_1$ in Figure 11.8). We have carefully considered the distribution of business travel and leisure travel. In total, there is currently around 53 per cent business travel at TOS. We have also considered the origin-destination (O-D) pattern, and calculated the total costs of the cheapest alternative transport mode for the relevant O-D pairs, including the time costs. These costs are then compared with the costs of using TOS for the same O-D pattern. By applying an average demand elasticity of -0.8, based on Gillen et al. (2003) and Helgheim (2002), the number of deterred passengers and hence the value of the shaded and crosshatched areas in Figure 11.8 can be computed.

In addition, the traffic effects from the MP shown in Figure 11.9, the O-D pairs and the alternative transport modes can be used for calculating the changes in emissions to air and the safety effects of using alternative transport modes. In most cases, the alternative mode will be to drive by car to an adjacent airport (BDU for most of the passengers) and go from there. We have also provided rough but careful estimates of the value of avoided delays when the airport reaches its capacity limits (assuming that 5 per cent of the passengers face 15 minutes of delay). Table 11.1 shows the results of the CBA for the MP.

Table 11.1 Economic effects of the Master Plan for Tromsø Airport

Economic effects		
Effect	**Discounted numbers (million NOK)**	
	Business	**Leisure**
Value of avoiding capacity problems, all travel	509	163
Sum all travel Whereof international journeys	*672* 23	
Avoided delays, passengers and airlines	30	
Safety (accident costs)	48	
Air emissions	- 66	
Investment costs for Avinor, MP on TOS	- 457	
Net present value (NPV) of the most important socio-economic effects	*227*	
Discounted revenue increase for Avinor, airside charges		
Increased revenue from MP for Avinor (million NOK)	280	

The extra travel costs of using adjacent airports for alternative routing are clearly the most significant cost component with a discounted value of around NOK 670 million. Accident costs are reduced (a positive benefit of the MP) because more road transport to adjacent airports is needed if the MP is not implemented. Emission costs increase (a negative benefit) because of increased aircraft movements due to the increased traffic ($X_0 - X_1$ in Figure 11.8) if the MP is implemented. In total, the MP for TOS appears to be profitable in a cost benefit perspective. It is worth noting that the discounted increase in revenues for Avinor shows a clear deficit (revenue + NOK 280 million, while the investments in the MP are NOK 457 million).

Because the alternative mode of transport without the MP measures has been air transport from adjacent airports, no significant extra maintenance and operation costs have been applied. However, if Bardufoss (BDU) as the nearest airport has to be used if TOS is not expanded, an extensive need for upgrading BDU is evident. These upgrading costs are estimated to be at least NOK 300 million, which strengthens the economic viability of TOS's Master Plan.

The network effects on the air transport system are not considered in detail. The main fraction of the market which will need to find other means of transport will have an additional loss as compared with what we have calculated. This would be the case because the departure frequency to relevant destinations would be lower on most of the adjacent airports. A weak positive effect might be that this transfer to adjacent airports would attract some new passengers living near these airports because the departure frequency would have to be somewhat increased over time. Another network effect is the costs that the airlines may face because their operations might become more expensive. It is likely that the network effects would strengthen the economic profitability of the Master Plan.

Chapter 4 deals with the economic impact of air transport in remoter regions, where employment and production multipliers are presented. We refer to this chapter for a more detailed description and discussion of multiplier effects and input-output models. For TOS, the employment multiplier is estimated to be 1.34 for employment and 1.44 for production (i.e. indirect and induced effects, Bråthen and Husdal, 2008). Given that the Master Plan for TOS should be implemented and the traffic predictions for the highest scenario would be met, then direct employment at the airport may increase from 1,020 man-years in 2006 to around 2,400 man-years in 2050. The sum of direct, indirect and induced employment may increase from 1,600 employees in 2006 to 3,800 employees in 2050. The total production value derived from the airport activities may increase from NOK 570 million in 2006 to around NOK 1,400 million in 2050. Even without any catalytic effects (described in Chapter 4 as presumably the most important derived effects of the air transport system), the impacts on the local economic activity level might be considered as quite substantial. These numbers are however uncertain because technological development in such a long time span may change the input-output coefficients in the model. It is also well worth noting that, like in other multiplier effect studies, these numbers are not net effects because without the Master Plan implemented, people who will not be employed in airport-related activities may find their jobs elsewhere.

Discussion and Conclusion

Timing of investments is important. If the capacity constraint should become effective from 2012 instead of 2015 as assumed in the CBA, the user benefits of having the capacity in place will increase by 60 per cent as compared with the numbers in Table 11.1. Conversely, if the capacity constraint should become effective from 2018 instead, then the user benefits will be reduced by 40 per cent. Larger investments that are made too early will have a particularly adverse effect on the economic profitability of the Master Plan. The importance of being able to fine-tune investments according to market needs is well worth noting.

The interesting main conclusion is that, from a financial perspective, the MP seems to be unprofitable. However seen from an economic point of view, where a broader picture of the use of real resources like passenger travel time is taken into consideration, the MP seems to be clearly profitable. This means that when the users' benefits of an air transport infrastructure improvement are taken into consideration, the result might be that a financially unprofitable project may prove to be viable in a socio-economic perspective. In general, in cases where there are economics of scale in production, and where the users are a part of the production function of the services (e.g. the passengers spend their own time in getting from A to B), then financial and economic appraisals might give different results. This observation might be of some importance where public air transport policy is concerned. This might be the case particularly where there are thin traffic markets which may not prove to be financially viable but nevertheless may well be economically profitable from society's point of view.

References

Bråthen, S., Eriksen, K.S., Hjelle, H.M. and Killi, M. (2000). Economic appraisal in Norwegian aviation. *Journal of Air Transport Management*, 6/3, 153–66.

Bråthen, S., Eriksen, K.S., Johansen, S., Killi, M., Lillebakk, L.M., Lyche, L., Sandvick, E.T., Strand, S. and Thune-Larsen, H. (2006). Samfunnsøkonomiske analyser innen luftfart. Samfunnsøkonomi og ringvirkninger. Del 1: Veileder. Report 0606a, Molde Research Institute/Institute of Transport Economics, Molde/Oslo.

Bråthen, S. and Husdal, J. (2008). En samfunnsmessig analyse av behovet for videreutvikling av Tromsø Lufthavn, Langnes. Report 0704, Molde Research Institute, Molde.

Gillen, D.W., Morrison, W.G. and Stewart, C. (2003). Air Travel Demand Elasticities: Concepts, Issues and Measurement. Department of Finance, Government of Canada.

Helgheim, B.I. (2002). Norsk luftfart. Avgifter og elastisiteter. MSc thesis in Logistics, Molde University College, Molde.

Lian, J.I., Rideng, A. and Denstadli, J.M. (2006). Reisevaner på fly 2005. Report 828/2006, Institute of Transport Economics, Oslo.

Ramjerdi, F. (1997). The Norwegian Value of Time Study. Part 1. Report 39/1997, Institute of Transport Economics, Oslo.

Thune-Larsen, H. (2006).Trafikkscenarier for Tromsø Lufthavn 2005–2020 og perspektiver fram mot 2050. Grunnlagsdokument for Masterplan 2004–2050. TØI, Arbeidsdokument ØL/1923/2006, Oslo.

Varian, H.R. (2006). *Intermediate Microeconomics: A Modern Approach*. 7th edn, W.W. Norton & Co, New York.

PART D
Innovation

Chapter 12
Virtual Airlines: The Concept of Local Heroes (Not) Flying

Lennart Bergbom and Johan Holmér

Defining a Virtual Airline

'Anyone can start their own airline with our assistance – it's easy as 1-2-3.' With this slogan Sverigeflyg, a Swedish virtual airline, attempts to market its business idea in Europe under the brand 'Wings of Europe' in order to establish a local or regional airline brand by using a basic concept: attracting loyalty and risk sharing from local business (Figure 12.1). So far Sverigeflyg has been able to establish a market presence in several domestic city pairs. But virtual airlines are not airlines by any traditional means. They hold no air operations licence, nor do they necessarily need to understand core airline economics. Instead they rent aircraft operations, i.e. aircraft, technical competence, maintenance, flight crew etc., through long-term subcontracts. Core business and competence is instead in marketing and ticketing, without head-to-head competition from traditional airlines. This has been proclaimed as a new cost-effective business concept, where substantial capital costs are lifted from the virtual airline's balance sheet to be part of a wet-lease contract deal. The question is whether the idea behind virtual airlines is as cost saving as one might predict. This chapter offers a brief examination of Sweden's experience of virtual airlines, including the business concept and market development.

Background

The history of virtual airlines in Sweden begins in 2001 when regional brand Gotlandsflyg was established by a group of local business entrepreneurs on the Baltic island of Gotland, reacting against high monopoly fares for regional air travel between Gotland and the Swedish capital, Stockholm. They claimed that this fare development was damaging the important island tourism industry and pushing unfair travel costs on to local businesses. For many years the 300,000+ annual passenger route had one sole operator to choose from (Skyways) and fares, both leisure and business, had seen a steady increase: between 1992 and 2001 the cost of a standard leisure ticket increased by 100 per cent and a standard coach fare by 80 per cent (real price adjusted). The high fares were also acknowledged by the Swedish Civil Aviation Administration (CAA), which stated in a report that the government should consider imposing public service obligation (PSO) on the route. The entrepreneurs asked local businesses, politicians and Gotland islanders

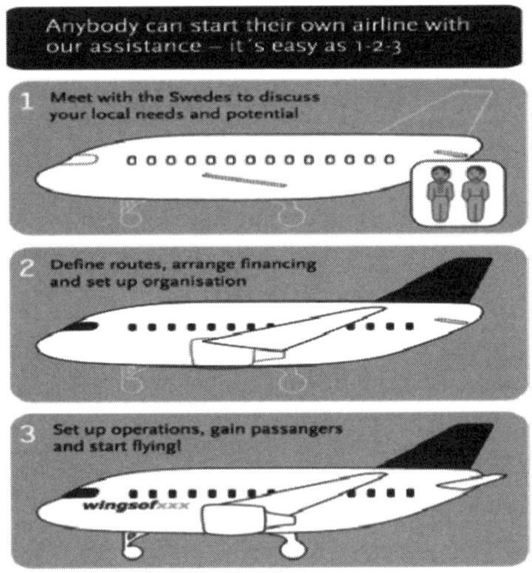

Figure 12.1 Wings of Europe business approach

Source: www.wingsofeurope.com

in general to support the creation of an independent Gotland airline. Shared risk capital and a loyal customer base committed to the airline from both the private and public sector gave steady grounds for a successful market launch.

Market Development: The Swedish Experience

The holding company, Sverigeflyg, has established six regional (virtual) airlines[1] and operates today with a €50 million turnover and a positive profit margin (Figure 12.2). In fact, nowadays Sverigeflyg is not merely flying under a wet-lease contract; it has recently purchased its own airline (Avitrans) that operates on some point-to-point routes within the Sverigeflyg Group. Although Sverigeflyg is the dominant virtual regional airline in Sweden there are other examples of local initiatives that have followed suit. Höga Kusten Flyg operates between Örnsköldsvik and Stockholm under a local initiative. Not all virtual airlines have experienced a successful market entry, though. Only a year after a high-profile market entry, Goodjet went bust. Goodjet had a different market strategy as it focused on head-to-head competition with the incumbent SAS and Malmo Aviation on Sweden´s major domestic routes, Stockholm–Malmö and Stockholm–Gothenburg. Various factors contributed to the enhancement of operations, but clearly the structure of production in itself was not enough to enter the market successfully.

1 Blekingeflyg (Ronneby), Gotlandsflyg (Visby), Kullaflyg (Helsingborg-Ängelholm), Flysmaland (Växjö), Sundsvallsflyg (Sundsvall) and Umeflyg (Umeå).

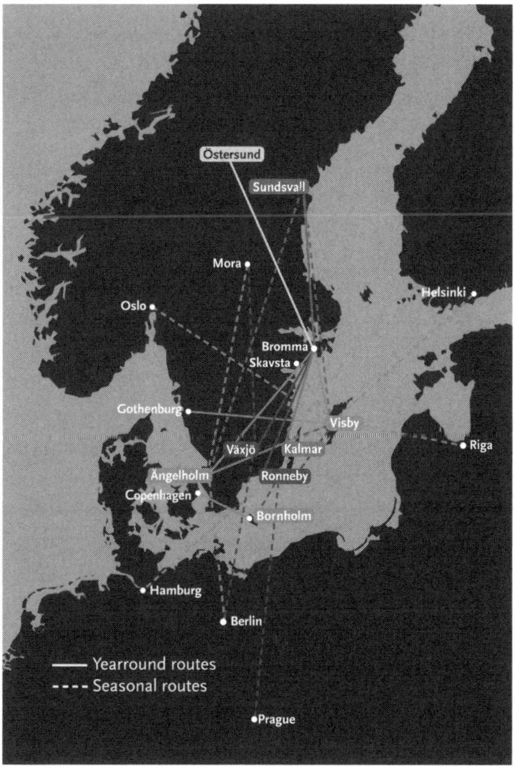

Figure 12.2 Virtual scope: Sverigeflyg holding company destinations in 2008

Source: www.wingsofeurope.com

Today's virtual airlines typically operate Fokker 50 and Saab 2000 turboprops with 50 seats on domestic sectors up to 400 km. According to Sverigeflyg the fleet concept is based on uncomplicated turnaround operations using turboprops, and operating costs (or aircraft wet-lease costs in this case) are kept low (according to Sverigeflyg) thanks to load factors averaging 75 per cent.

The Economics of Virtual Airlines

The essence of a virtual airline business model is founded on some basic factors. The information given overleaf is based on evidence given by virtual airlines operating in Sweden.

- *Local ownership*. The virtual airlines are jointly owned together with local business interests. Besides capital risk sharing, local investors mean a loyal customer base that could be widened via local business-to-business contacts. Furthermore, there is local insight regarding demand for potential new air services from the region and service quality, including timetables and early knowledge of whether customers are happy with the service concept. What works in one city and region never works fully in another; therefore local know-how is invaluable, according to Sverigeflyg.
- *Local management*. Local management knows the local needs, has the network necessary to maximize sales and generate client feedback and the local structure gives the ability to move fast and make changes when needed. At the same time, by providing central support in a number of areas where economies of scale are present, as is the case with Sverigeflyg, cost-effective measures can be taken. This is evident in areas such as budgeting, scheduling, organizational development, staff recruitment/development and retention etc.
- *Point-to-point operations*. Most regional Swedish carriers typically tend to act, at least partially, as feeders to major network carriers. The experience in Sweden so far shows that virtual airlines base their concept on direct services without interlining or code-sharing arrangements.
- *Inflight services*. Besides management and sales positions, the virtual airlines have focused on inflight image by employing their own cabin crew in an otherwise wet-lease operation.
- Finally, they have a cost base without aircraft, cockpit crew and maintenance.

Holding Your Own AOC versus Using a Partner Airline

There are a number of ways to finance aircraft for scheduled passenger traffic. Major airlines normally use a mix of aircraft ownership, dry leases and wet leases. An airline that owns its fleet and wishes to reduce its capacity can of course sell aircraft to other airlines, but there is always a risk involved, due to the volatility of aircraft values. That risk can be reduced to the risk preference of the airline by choosing from leasing contracts with different durations and conditions; but, in an efficient market, the purchase of risk reduction will always come at a cost. The point here is that airlines can trade risk against expected cost using almost a continuum of financial services ranging from ownership to short-term leases.

The virtual airline cannot, by definition, make direct use of any aircraft as it cannot hold an Air Operator's Certificate (AOC). Instead, the virtual airline relies on contracts with one or several AOC holders. The relation between the virtual airline and the AOC holder(s) can in principle be short, mid or long term, even though one major benefit of being a virtual airline is flexibility, which speaks in favour of short- to medium-term contracts.

Outsourcing Operations to One or More AOC Holders

The largest virtual airline in Sweden, Sverigeflyg Holding AB, can serve as an example illustrating some of the aspects involved in the relations between a virtual airline and its flight provider(s). Operations were started in April 2001 between Visby on the Swedish island of Gotland and Stockholm-Bromma Airport.[2] The service was marketed under the brand Gotlandsflyg but, apart from the first 16 days, has been operated by the airlines Trygg-Flyg AB (April 2001–February 2002), Golden Air AB (February 2002–June 2007) and Skyways Express AB (March 2007 onwards).[3] The network has since been widened step-by-step and consists of ten routes operated all year and eight seasonal routes.[4]

According to the airline, the purpose of the arrangement described above is:

- to run the operations under competitive pressure in order to minimize operational costs;
- to limit cost unpredictability by reducing operational, maintenance and labour market risks;
- to make it simple and cost effective to reduce capacity quickly, if that is needed.[5]

By using independent airlines for the flight operations, the virtual airline can use competitive pressure to increase its economic efficiency. First, there is an intertemporal competitive pressure, since these agreements will be renegotiated in a not too distant future. Second, the virtual airline can choose to use a tendering process, which will incur competitive pressure between different contenders/airlines. In addition to the competitive aspects, the overall setup is similar to a charter agreement, where the risks of unforeseen maintenance costs, labour market conflicts etc. are completely borne by the airline during the contracting period. Finally, the contracts are designed to make it quick and simple for the virtual airline to reduce costs by downsizing operations. The last point may be a key issue, as many airlines have experienced difficulties in limiting losses during downturns in the business cycle. Since the start of Gotlandsflyg/Sverigeflyg in 2001, the Swedish economy has been in good shape, meaning that the first real test of the business model will be the current worldwide economic crisis.

The obvious cost of this arrangement for the virtual airline is that it has to pay a premium to shift the risks from itself to the airline company. In the case of Sverigeflyg, the risk reduction is said to be worth the premium, since lack of

2 Gotlandsflyg AB, annual report 2001.

3 Traffic statistics, The LFV Group.

4 <http://www.sverigeflyg.se>, 25 August 2008.

5 Interview with Daniel Långsten, CEO at Wings of Europe, a development company within the Sverigeflyg Group, 14 August 2008.

flexibility has led to financial problems for many airlines.[6] In addition to this, the financial record of the company supports the view that the virtual airline setup may be a significant positive factor for the company.

Combining External Flight Providers with Airline Ownership

Sverigeflyg has recently developed its operational model through the purchase of the airline Avitrans Nordic AB in 2007.[7] This could be interpreted as a move by Sverigeflyg towards traditional airline thinking, where in-house flight capacity is combined with external flight operators. The reason for using such a setup is normally to add marginal flexibility to a production model, which is based on aircraft either owned or leased on a long-term basis.

In this case, however, the reasoning is different. The purpose behind the purchase of Avitrans Nordic AB was to 'get control over the local [airline] companies' slots at Bromma [airport]'.[8] In contrast to a traditional airline, Sverigeflyg could not control the slots of its own route network. Instead, the flight operators controlled the slots, since slots can only be held by airlines. This made it difficult for Sverigeflyg to change operators freely, as it could be almost impossible to secure slots for a new airline entering an airport where suitable slots are already in use. The solution was to buy an airline, which made it possible to start getting control over slots.

Another related point is that the slots that Avitrans control can be used by another flight provider in a wet-lease agreement between Avitrans and an external flight operator, without Avitrans losing its grandfather rights to the slots in question. The purpose of this model for Sverigeflyg is not to transfer all, or even a majority, of its flights to its own airline, Avitrans, but to continue with a production model where Sverigeflyg can combine a large share of external flight operators with control of attractive slots.

The reason for keeping a large share of production outside its own airline, Avitrans, is the ability for Sverigeflyg to continue its strict cost control, where flight providers are frequently evaluated in a competitive process to optimize cost efficiency.

The Financial Performance of Sverigeflyg

In contrast to many airlines Sverigeflyg does not report its passenger figures. In the case of Sverigeflyg the traffic growth has been so rapid in relation to the air fare volatility that turnover can be used as a rather accurate proxy for traffic volume. Figure 12.3 clearly illustrates the high and consistent growth of Sverigeflyg since the start of the airline. As can be seen, the growth consists of two parts: first, the

6 Interview with Daniel Långsten, 14 August 2008.

7 Avitrans Nordic AB, annual report 2007.

8 Sverigeflyg Holding AB, annual report 2007, p. 3.

two major bases, Visby (Gotlandsflyg) and Ängelholm-Helsingborg (Kullaflyg), have grown; and, second, more markets have been added to the network.

Sverigeflyg is clearly a success in terms of volume but, in relation to the overall traffic development of Swedish domestic aviation, the success is even larger. The volume growth of Sverigeflyg has been 62 per cent a year on average from 2001 to 2007, whereas the number of passengers in Swedish domestic traffic has decreased by 2.2 per cent per year on average over the same period.

Return on Investment

In the history of aviation many airlines have been able to create traffic growth but it has been difficult for the industry to achieve sustainable profits over longer periods. Sverigeflyg Holding AB was founded almost two years after the start of Gotlandsflyg. Furthermore, from 2007 the airline Avitrans Nordic AB has been a significant part of the company group. Therefore, the calculation of profits and losses below is based on the annual reports of the virtual airline subsidiaries, starting with Gotlandsflyg AB in 2001 and excluding Avitrans Nordic AB. With this approach a time series as consistent as possible is created. Nevertheless, the reported performance of Sverigeflyg Holding AB is used for return on equity as these data more accurately reflect the actual return to the shareholders.

Table 12.1 clearly shows that Gotlandsflyg and Kullaflyg both experienced first-year losses but that they were soon turned into consistent profits. With the exception of Flysmaland the performance of the virtual airlines started from 2005 onwards has been significantly weaker. The overall trend is, though, an increase in profit of the Sverigeflyg Group.

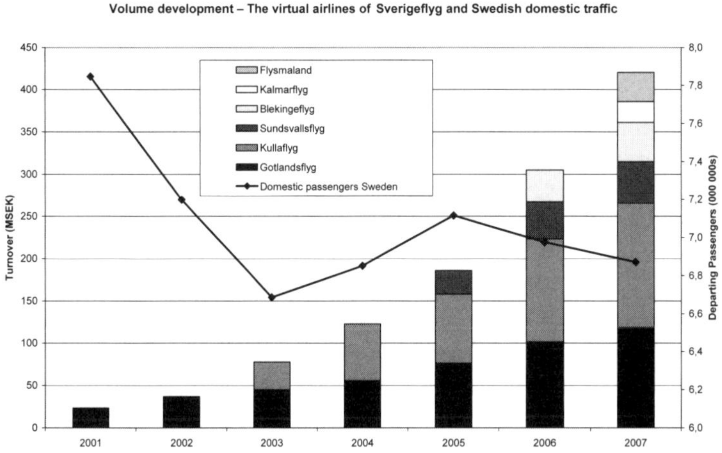

Figure 12.3 Sverigeflyg revenue and Swedish domestic traffic

Source: Annual reports, Sverigeflyg AB and Swedish CAA

Table 12.1 Pre-tax profit of the Sverigeflyg virtual airlines (million SEK)

	2001	2002	2003	2004	2005	2006	2007	2001–07
Gotlandsflyg	-3.3	3.0	2.6	3.8	8.1	7.7	12.5	34.5
Kullaflyg			-4.5	6.8	10.4	14.9	23.2	50.8
Sundsvallsflyg					-8.6	-0.5	0.1	-8.9
Blekingeflyg						0.4	-1.6	-1.2
Kalmarflyg							-5.0	-5.0
Flysmaland							4.0	4.0
Total	-3.3	3.0	-1.9	10.6	9.9	22.5	33.2	74.2
% Equity return	-294	103	-5	69	31	62	29	25

Source: Annual reports, Sverigeflyg AB

Return on equity is calculated as profit after tax divided by the average amount of equity during the calendar year in question. It has clearly varied over the years but the average yearly performance of 25 per cent return on equity must be seen as evidence of a business model that so far has been able to create excellent returns to the owners, especially when the performance is compared with other regional airline activities in Sweden and in many other countries.

Passenger Rights on Virtual Airlines

In 2004, the European Union (EU) adopted EC Regulation 261/2004, thereby creating transparent and extensive airline passenger rights that primarily required airlines to compensate air travellers when flights were delayed and cancelled. The new legislation on passenger rights wiped out any previous doubts on what protection airline passengers could expect and from whom compensation should be claimed. Still, in some cases it might be difficult for the air passenger to identify the operating air carrier at the time of ticket purchase when flying with a virtual airline. Before Regulation 261 entered into force in February 2005, as virtual airlines began their market entry, Sweden therefore considered a number of guidelines and regulations in order to grasp the requirement to better inform passengers on flight information and compensatory rights. This never happened but current EU legislation, with the obligation to inform passengers of their rights, seems to work satisfactorily in this matter.

Conclusions

To conclude it would be fair to say that the concept of virtual airlines is not very much of an airline business phenomenon since most of its concept can be found among traditional air carriers. Outsourcing of operations, maintenance and labour is today mainstream strategy. What is interesting however is the degree of successful market entry combined with a positive economic performance in terms of return on investment. This is why the core concept, serving regional point-to-point routes and sharing the risk together with local business, leads to the assumption that the prospects for continued market entry in Sweden and elsewhere looks promising for the virtual airline industry.

Chapter 13
Transportation of Goods and Passengers to Remote Areas Using Airships: Two Case Studies in India

Rajkumar S. Pant

Background

In the span of about a century, mankind has gone from making a few test flights in earth's atmosphere to orbiting celestial bodies in outer space, from sliding across sand dunes to spanning even the mightiest of the oceans. Where on one hand man has bridged the time and distance gap with the advent of the jet age, critical issues like the environmental effects of growth in aviation and rising fuel prices are putting a question mark on the future of conventional modes of aviation. As a result, the growth rate of civil aviation is reaching a plateau, and is expected to experience a slowdown in the near future.

Aerospace engineers from across the world are now focusing their attention and efforts to come up with new innovative solutions, and also to revive technologies once considered obsolete, in order to address the aerial transportation needs of the general public across the world. Developing cost-effective transportation systems for air travel has become increasingly important, since congestion on surface modes of transport has already reached horrendous levels.

Provision of air connectivity to communities in remote locations usually comes with a huge price tag, mainly due to the high cost of setting up and maintaining infrastructure for a relatively small number of travellers. One solution to this problem that is catching the attention of researchers and potential investors worldwide is airships, which are aerial vehicles based on lighter-than-air (LTA) technology.

During the 1930s, airships were luxury liners of the sky, ferrying passengers across the Atlantic in comfort and style, a far cry from the cramped conditions that a vast majority of air travellers today are forced to put up with. Today, however, in the minds of the general public, the very word "airship" conjures up a vision of a bygone era, and airships suffer from a connotation of being dangerous and unsafe. This is mainly due to a series of fatal accidents in the late thirties, which led to their downfall.

However, it is being increasingly felt that the potential of these titanic floating giants that once roamed the skies has been hugely undermined. Recent developments in airship technology have made airships much safer, chiefly due to the availability of helium as an inert LTA gas, and far superior material and control systems. Many researchers today believe that airships present an effective, low-cost, environmentally friendly solution for some niche areas of air transportation. This has spawned many studies and technology development initiatives worldwide, and it can be said that airships are now undergoing a revival phase.

Airships can be used very effectively in regions where economic considerations are a key driver towards solving transportation problems, especially in providing air transport services to remote communities. The costs related to setting up and operating the infrastructure required for operating airships are quite small compared to those for their heavier-than-air counterparts. Further, the operating costs of airships can also be quite low, primarily due to their low fuel consumption.

Introduction to Airships

Airships are LTA vehicles that work on Archimedes' principle, i.e. "Bodies submerged into a fluid receive from it a lifting force which is equal to the mass of the displaced fluid"[1]. An airship's envelope is filled with a gas having a lower specific weight than the atmospheric gas; hence it generates lift, which is equal to the difference between the weight of displaced air and that of the lifting gas. This is called *static lift*, since it is available even when an airship is stationary. In addition to this, an airship also derives some *dynamic lift*, due to the action of aerodynamic forces acting on it as it moves through the air, just like an aeroplane.

Although airships can take off vertically, a higher payload capacity can be achieved by making a running takeoff in an airship, much like an aeroplane. The speed gained on the ground is converted to lift when the pilot raises the airship's nose. Once airborne, airships can perform much like helicopters, remaining nearly geostationary for extended periods of time, albeit with much less fuel consumption and noise and vibration levels.

The reason for much lower fuel consumption of airships is quite apparent; a large fraction of a heavier-than-air vehicle's propulsive power is spent in achieving enough relative velocity between the vehicle and ambient air to overcome gravity. In LTA vehicles, gravity can be overcome purely by static lift, even at zero forward speed.

Types of Airships

Based on their internal structure, airships can be classified into three categories, viz. rigid, semi-rigid and non-rigid, as shown in Figure 13.1.

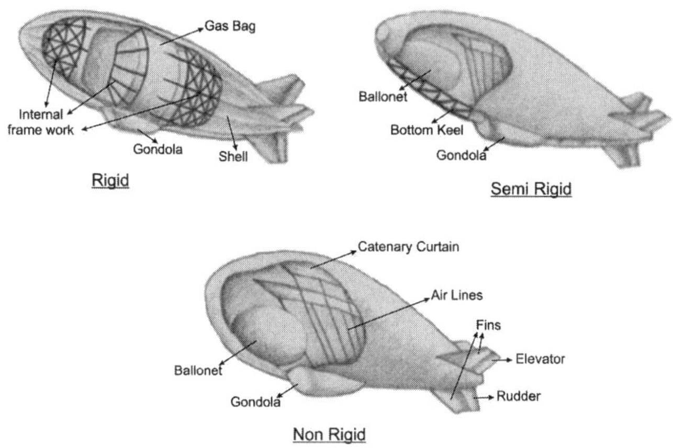

Figure 13.1 The three types of structural layout of airships

Source: adapted from Anderson, J. D., *Fundamentals of Aerodynamics* [2].

A rigid airship is one in which the envelope retains its shape by the use of an external rigid shell, with an internal structural framework consisting of transverse girders and longitudinal frames. The LTA gas is carried in several bags which are located within the transverse frames. Rigid airships are found to be structurally efficient only when their sizes are very large, with envelope volume in excess of 30,000 m^3 [3]. Rigid airships were produced and successfully deployed from the beginning of the 1900s to the end of the 1930s, but are now considered obsolete; their heyday ended when *The Hindenburg* was destroyed on 6 May 1937. The need for an internal framework has been overcome by making use of highly durable load-bearing material. Improvised manufacturing processes also enable airship manufacturers to harness the potential of high-quality materials.

Semi-rigid airships have a non-rigid envelope supported by an internal framework. These often consist of a rigid (but occasionally flexible) keel frame along the long axis under the aerodynamic hull envelope. The framework of the airship is responsible for the distribution of the suspension loads of the various attachments and the lifting gas loads evenly throughout the hull's surface. It is also used to partially relieve stresses on the hull during manoeuvres.

A non-rigid airship, or "blimp" as it is colloquially called, is an airship without an internal supporting framework or keel. A non-rigid airship differs from a semi-rigid and a rigid airship in that it does not have any rigid structure, neither a complete framework nor a partial keel, to help the airbag maintain its shape. Rather, these aircraft rely on the load-bearing capacity of a flexible envelope material, and the shape of the envelope is maintained solely by the internal overpressure of the lighter-than-air gas. The only solid parts are the passenger car and the tail fins.

Most of the operational airships today are of the non-rigid type, mainly due to the developments in materials technology, enabling the entire load to be carried by the envelope material itself. Another point in their favour is ease in transportability; their envelopes can be deflated and packed in small containers and shipped. A major disadvantage of rigid and semi-rigid airships is that they have to be flown to the point of delivery since they cannot be easily packed in small size containers, like non-rigid airships can. However, semi-rigid airships are far superior to non-rigid types in structural efficiency, and offer more scope to designers to come up with innovative layouts, since the concentrated loads acting on the vehicle can be transferred to the structural elements.

Capabilities and Limitations of Airships

Some of the advantages of airships as compared to conventional modes of transport are listed below:

- Since airships are lighter-than-air vehicles they do not use energy for lift, but only for forward movement. Also, unlike conventional modes of transport, airships are relatively quiet and have extraordinary endurance (they can fly for weeks instead of hours) compared to conventional aircraft.
- Airships do not need runways to take off and land, hence their operating bases can be located away from the congested airspace near an existing airport. With the use of a high mooring mast, their operation from or near city centres is technically feasible. Further, airships fly at lower altitudes (around 2,000 ft above ground level). In fact, any increase in operating altitude adversely affects their payload carrying capacity. Hence, the introduction of airships for transportation of goods and passengers does not interfere too much with the existing airspace in which conventional aircraft operate, and also does not add to congestion in an existing airspace.
- Airships can be configured to perform vertical lift operations with much lower fuel consumption than helicopters.
- Since airships operate at low altitudes, and also climb and descend gradually, it is not necessary to pressurise the passenger cabin; hence it can be made spacious and luxurious. Airships also offer a breathtaking view of the outside world for the passengers.
- Owing to their low speed of operation and inherent flexibility in their structure, airships are safer than aircraft in handling operating hazards such as bird hits, or accidental hits to fixed objects. Helium is now available globally in fairly large quantities; hence its use as a lighter-than-air gas can make airships a very safe mode of air travel.
- The two main aspects of civil aviation that harm the environment are emissions and noise, both of which are much lower in airships, compared to that required by the equivalent conventional aircraft. This is primarily due to the fact that the propulsive power (hence engine size and/or fuel consumption) required by airships is much smaller.

- Airships can achieve level flight speeds of up to 100 km/hr. Hence, they can cover a distance of 500 km in around four to five hours. The operating cost per seat-km of an airship can be far lower than that of a conventional aircraft, thus giving it a competitive edge.

However, airships do suffer from some operating limitations as compared to conventional modes of transport, some of which are listed below:

- Airships are sensitive to changes in ambient wind speed; hence they are not *all-weather* vehicles. Airship operation can be hindered in disturbed weather conditions, especially when accompanied by vigourous changes in the direction and magnitude of ambient winds, affecting their dispatch reliability.
- Due to their large size and susceptibility to wind disturbance, airships need to be stored in large hangers.
- The payload carrying capacity of an airship diminishes a lot with increase in operating altitude.

Airships: Making a Comeback in the 21st Century

Some of the limitations of airships mentioned above can be overcome by developing hybrid configurations, especially for the heavy lift application. Liao and Pasternak [4] have carried out a detailed review of current airship development programmes worldwide, as well as techniques for structural analysis, modelling and simulation.

With growing environmental concerns and with the use of a safe gas like helium instead of hydrogen, aerospace engineers are now looking at reviving the antiquated technology of airships. But still, their numbers worldwide are quite low; by the beginning of 2000, there were around 29 airships that were operational in Argentina, Brazil, China, Germany, Hong Kong, Italy, Russia, Singapore, the UK and the USA.

Zeppelin Luftschifftechnik GmbH, based in Friedrichshafen, Germany, is the successor of Luftschiffsbau-Zeppelin, which had earlier pioneered passenger airships at the beginning of the 20th century. This company has revived transport airships in the early 1990s, by developing Zeppelin NT class semi-rigid passenger airships, which are the most successful passenger carrying airship of recent times. Zeppelin NT, which has a semi-rigid structure, consists of a keel on the bottom and the envelope supported by carbon-fibre frames at various locations. This layout enables the engines to be mounted on the side and back of the envelope, thus providing much higher manoeuvrability and huge reduction in the manpower required for ground operations.

Of late, several countries are actively developing and operating passenger airships for various applications. US LTA Corporation and American Blimp

Corporation have taken the lead in manufacturing airships for both civil and military applications. The new airships developed by these firms feature a gondola capable of carrying nine passengers plus the pilot, and endurance exceeding 24 hours. Airships are presently being used in the southern parts of USA such as Florida for aerial sightseeing and tourism. Owing to the success of airship operations for tourism development, serious thought has been given to short-distance commuting using airships.

With growing awareness of rising fuel costs, leisure travellers from across the world are now willing to accept newer and more environmentally friendly means of transport. Industries in major Asian countries such as China and India are now investing some of their time and resources towards the revival of the airship. In 2004, Hua Jiao Airship China, a private firm engaged in commercial airship services, received flying licences for its airships from the Civil Aviation Administration of China (CAAC). This company is the first and only manufacturer of airships in Asia.

Airships for Providing Air Connectivity to Remote Areas

In India, the Program on Airship Design and Development (PADD) was initiated in the early years of the new millennium. PADD aims at developing indigenous airship technology in India for various scientific and commercial applications. In the first phase of PADD, conceptual design studies of airships were carried out for transportation of goods and passengers in a remote region in Uttarakhand (earlier known as Uttaranchal), a province of north India, over mountainous terrain under 'hot and high' conditions. Techno-economic feasibility of leasing airships for operation in India was investigated, and a Project Definition Report for the development of two airships, the PADD *Demo* and the PADD *PaxCargo*, was completed in 2003 [5].

Two case studies were also carried out to bring out the efficacy of airships as an alternative means of transportation, details of which are presented in the sections that follow. In the first case study [6], a techno-economical comparison of helicopters and airships for ferrying passengers from low-lying plains to mountainous regions was carried out. The direct operating cost (DOC) and payload capability of two five-seater non-rigid airships and two five-seater helicopters were compared for connecting three specific city-pairs in Uttarakhand.

The second case study [7] looked at operation of airships for a popular pilgrimage route in the same area, as against the current mode involving a large amount of travel by road vehicles and trekking. A cost-benefit analysis of operating costs was made between a 13-seater airship and two road vehicles, a three-seater luxury car (limousine) and a seven-seater sports utility vehicle (SUV). Details of these case studies are provided in the next section.

A Brief Introduction to Uttarakhand

Uttarakhand is a relatively new province in north India, which was carved out of the erstwhile large province named Uttar Pradesh in 2000. It is located in the Himalayan region with many important cities, hill stations and scenic locations. A large fraction of revenue in Uttarakhand is generated by tourism. However, the majority of the land in Uttarakhand is mountainous and difficult to navigate. Added to this, the condition and quality of the existing road network is poor, and many major roads are badly maintained. Roadblocks are very common due to sudden landslides. Due to the terrain features, the rail network is limited to the lower reaches of Uttarakhand. Journeys of only 100 km can take up to four hours, sometimes even longer, depending on the weather conditions. This causes frequent traffic bottlenecks and delays, leading to the need for an alternative mode of transportation that does not depend on existing road or rail networks. Aircraft are ruled out since the terrain features preclude the possibility of building airports. Hence, a suitable mode of transport had to be identified to carry passengers across remote regions of Uttarakhand.

Case Study I: Comparative Analysis of Five-Seater Airships and Helicopters

Airships and helicopters are the only possible forms of aerial vehicles that can circumvent some of the difficulties mentioned above. Compared to fixed-wing aircraft, they can be operated over mountainous terrains more safely, since they can be flown at much lower speeds. Thus, they can fly through relatively narrower valleys or corridors with higher levels of tolerance and safety since their flight dynamics do not require large forward velocity as their prime source of lift. They also require significantly less area for landing, parking and taxiing, which makes them ideally suited for operation in mountains, where it is impossible to construct regular runways due to the unavailability of large areas of flat land. However, increase in temperature and operating altitude adversely affect the performance capability and payload carrying capacity of airships and helicopters much more than for fixed-wing aircraft. Compared to helicopters, airships have significantly lower fuel consumption, higher endurance and offer a much better ride quality to the occupants. Airships, on the other hand, are more sensitive to weather fluctuations, and are larger in dimensions compared to helicopters of similar capacity.

Selection of Airships and Helicopters

Table 13.1 lists the values of a few important parameters of the two five-seater, non-rigid airships and two five-seater helicopters that were considered for this study. The data related to the airships were obtained from [8] and [9], and to the helicopters from [10].

Table 13.1 Important parameters of the selected airships and helicopters

	5-seater airships		5-seater helicopters	
Description	**TLG A60+**	**US-LTA 138S**	**Bell 206L-4**	**Bell 407**
Envelope volume (m^3)/ Rotor diameter (m)	1,926	3,908	11.28	10.68
Empty weight (kg)	1,300	2,673	1,056	1,210
Max. takeoff weight (kg)	2,000	4,213	2,064	2,268
Max. length (m)	39.6	48.8	12.91	12.61
Width (m)	11.0	12.7	2.33	2.47
Height (m)	13.4	17.3	3.04	3.10
Cruise ceiling (m)	2,225	2,743	2,356	4,359

Selection of the Three Routes

After studying the geology and topology of Uttarakhand, two levels of operation for airships and helicopters were considered, the lower level comprising all locations below an altitude of 2000 m and the upper level consisting of locations situated above this altitude. An analysis of the population distribution of Uttarakhand revealed that 7,054,700 people live in the lower level, which constitutes nearly 68 per cent of the total population [11]. Further, a majority of agricultural and commercial activity takes place in this level. Locations in the higher level were beyond the scope of operation of existing airships and helicopters; hence this study was limited to operation only in the lower level.

Based on a study of demographic data of Uttarakhand [11], three city-pairs were chosen for this comparative study, viz. Dehradun to Nainital, Pauri to Almora and Uttarkashi to Pithoragarh. A feasibility study on operation of airships between these cities in Uttarakhand was carried out by The Lightship Group [12], and the exact routes to be followed between the three major city-pairs on which the airships were to be flown were identified, as shown in Figure 13.2. It was assumed that the helicopters would also be operated along the same routes.

Operating Costs

Operating costs are applicable to all aircraft, and are often a decisive factor when evaluating the relative merits of certain aircraft having comparable performance characteristics. The factors that contributed to direct operating cost (DOC), such as depreciation and insurance, maintenance (labour and spares), crew remuneration and fuel costs, have been considered. A description of the methodology adopted for estimation of DOC due to these factors follows.

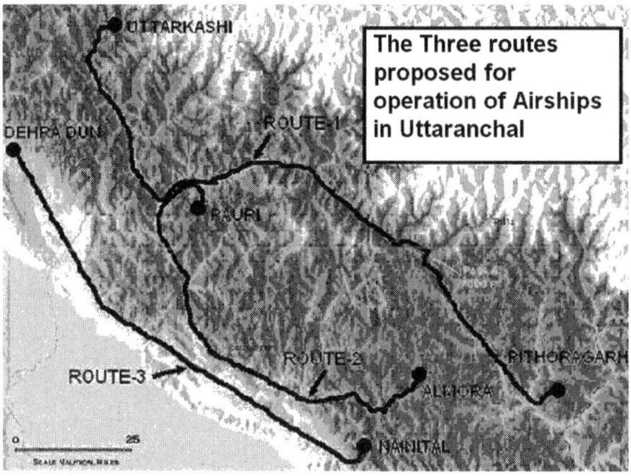

Figure 13.2 Map of Uttarakhand highlighting the three routes for airship and helicopter operation [12]

Depreciation and insurance Annual cost of depreciation was taken as 10 per cent of the purchase price of the airship or helicopter. Annual insurance premium has been taken as 2 per cent of the insured value of the airship/helicopter and support equipment, as suggested in a previous study [7].

Maintenance The annual cost for labour and spares for maintenance of the US-LTA 138S airship was obtained from [8]. These costs were scaled as a percentage of total procurement cost for TLG A60+ also. The maintenance schedule (in terms of maintenance man hours per flight hour, or MMH/FH) and cost figures for the helicopters were obtained from [10]. The results of the same have been summarised in Table 13.2.

Table 13.2 Annual costs of maintenance for the selected airships and helicopters

Name	MMH/FH	AMT[2] (in hr)	Annual cost[1] (in million US$)	
			Labour[3]	Spares
TLG A60+			0.051	0.090
US-LTA 138S			0.075	0.132
Bell 206L-4	0.841	1100.03	0.072	0.249
Bell 407	1.190	1333.99	0.087	0.280

Notes: [1] Assuming annual usage for eight months. [2] AMT stands for annual maintenance time. [3] Labour charges assumed to be US$65 per hour [10].

Fuel expenditure Table 13.3 shows the results obtained on the three routes for the airships and helicopters. Since potentially hazardous weather conditions can pose a serious threat to airship operations for approximately three months of the year, the calculations are based on operation for eight months, keeping one extra month as a contingency time margin for major maintenance and spillovers. The daily and annual totals of flight time and cost of fuel over the three routes are shown in Table 13.4.

Table 13.3 Block time and block fuel results for the airships and helicopters

Name	Route	Speed (km/hr)	Range (km)	Time (hr)	Fuel (kg)
TLG A60+ (5-seater airship)	Dehradun to Nainital	84	182	2.67	51.73
	Pauri to Almora		204	2.93	56.81
	Uttarkashi to Pithoragarh		315	4.25	82.45
US-LTA 138S (5-seater airship)	Dehradun to Nainital	72	182	3.03	82.36
	Pauri to Almora		204	3.33	90.67
	Uttarkashi to Pithoragarh		315	4.88	132.60
Bell 206L-4 (5-seater helicopter)	Dehradun to Nainital	178.5	182	1.52	167.54
	Pauri to Almora		204	1.64	181.13
	Uttarkashi to Pithoragarh		315	2.26	249.68
Bell 407 (5-seater helicopter)	Dehradun to Nainital	222.5	182	1.32	170.02
	Pauri to Almora		204	1.42	182.77
	Uttarkashi to Pithoragarh		315	1.92	247.13

Table 13.4 Flight time and fuel cost estimates for the airships and helicopters

Airship	Daily total			Annual total (8 months)		
	Time	Fuel (kg)	Fuel cost (US$)	Time (hr)	Fuel (kg)	Fuel cost (million US$)
TLG A60+	09 h 51 min	191	538.02	2373	46,037	0.130
US-LTA 138S	11 h 17 min	306	860.91	2708	73,658	0.208
Bell 206L-4	05 h 26 min	598	443.22	1308	144,207	0.107
Bell 407	04 h 39 min	600	444.39	1121	144,609	0.108

It is interesting to note that although the helicopters can cover the routes in nearly half the time, their fuel consumption is around double or triple. However, helicopters use Jet-A fuel, which costs US$0.6 per litre, as compared to airships, which use 100 LL aviation fuel, which costs US$2.0 per litre. Hence, the annual fuel costs of helicopters are much lower, compared to airships!

Crew compensation It was assumed that the operations would require three pilots and one trainee, ten technicians and ground crew, and four sales and marketing personnel. Using the values quoted for a Skyship 600 airship modified for operation in Indian conditions [7], the crew compensation was assumed to be US$0.4 million per year, and the same was assumed to be applicable for both the airships, as well as the helicopters.

Comparison of DOC of Airships and Helicopters

Table 13.5 lists the breakdown of direct operating costs (DOC) of airships and helicopters. Helicopters are seen to have nearly the same annual DOC as the airships, but their hourly DOC is 75–150 percent higher.

Table 13.5 Estimation of DOC for two 5-seater airships and two 5-seater helicopters (million US$)

Cost breakdown	TLG A60+	US-LTA 138S	Bell 206L-4	Bell 407
Acquisition cost	2.100	3.338	1.640	1.910
Utilisation per month (hours)	296	338	163	140
Annual insurance premium	0.042	0.067	0.033	0.038
Annual fuel cost	0.130	0.208	0.107	0.108
Annual maintenance	0.051	0.075	0.072	0.087
Annual spares	0.090	0.132	0.249	0.280
Depreciation of hull and ground support infrastructure	0.210	0.334	0.164	0.191
Annual DOC (8 months)	0.923	1.216	1.025	1.104
Hourly DOC (US$)	388.96	448.88	783.49	985.01

Results

It was observed that the annual operating costs of airships and helicopters are in fact quite similar. But the hourly DOC, as well as hourly cost per seat for airships, is significantly lower than the corresponding figures for helicopters, as seen in Table 13.6.

Table 13.6 Comparison of DOC for airships and helicopters

Name	Annual usage (hr)	Annual DOC (million US$)	Annual cost per seat (million US$)	Hourly DOC (US$)	Hourly cost per seat (US$)
US-LTA 138S	2373	0.923	0.185	388.96	77.8
TLG A60+	2708	1.216	0.243	448.88	89.8
Bell 407	1308	1.025	0.205	783.49	156.7
Bell 206L-4	1121	1.104	0.221	985.01	197.0

It can be summarised that since the fuel consumption of airships is 1.7 to 2.7 times lower compared to helicopters, they provide a greener solution to meet transportation requirements. However, the flight duration by helicopters is nearly half compared to airships. Though the operating costs of airships and helicopters are quite similar, airships offer far superior levels of comfort due to larger cabin space and lower vibration levels, perhaps resulting in a more enjoyable flight, especially for tourists.

Case Study II: Char Dham Yatra – Pilgrimage in the Indian Himalayas

The "Char Dham Yatra", which means "the trip to four abodes or seats", is one of the most important Hindu pilgrimage circuits in the Indian Himalayas [13]. Located in the Garhwal section of the state of Uttarakhand, the circuit consists of four sites – Yamunotri, Gangotri, Kedarnath and Badrinath. While each of these sites is unique in its own fashion, their collective inclusion in the Char Dham Yatra has, over time, caused them to be viewed together in popular imagination and actual pilgrimage practice.

Every year, millions of pilgrims travel from different parts of the country to visit the holy pilgrimage centre located in the Indian Himalayas. Most pilgrims to the Char Dham set out from the famous temple town of Haridwar. Others leave from Haridwar's sister city, Rishikesh, or from Dehradun, the capital of Uttarakhand. From there, the tradition is to visit the temples and holy places of Yamunotri, Gangotri, Kedarnath and Badrinath in the respective order as mentioned. A road map of the Char Dham Yatra is shown in Figure 13.3.

1. Yamunotri, the source of the Yamuna River and the seat of the goddess Yamuna, is approximately 223 km from Rishikesh, Haridwar or Dehradun. The actual temple is only accessible by a 13 km trek from the town of Hanuman Chatti.

2. Gangotri, the source of the Ganga (Ganges) River, is located around 400 km from Rishikesh, Haridwar or Dehradun, or two days from Yamunotri. Pilgrims make an overnight 17 km trek to Gaumukh, which is the current source of the Ganga.
3. Kedarnath is approximately 341 km from either Gangotri or one of the main disembarkation points on the plains. The actual temple is accessible only after a steep 14 km walk.
4. Badrinath is a two-day journey covering approximately 207 km from one of the main disembarkation points on the plains. By far the most important of the four Char Dham sites, Badrinath receives many more visitors than the other three sites. The temple and its substantial surrounding village are accessible by road.

Since a majority of the pilgrims visiting these holy places are usually middle-aged or elderly individuals, most of them have to rely on the local modes of transport that are available. These transportation modes mainly include palanquins, animal-driven transport or conventional road transport. Since none of these are very comfortable means of commuting, many pilgrims choose to walk their way up to the temple premises. In order to provide a better and more comfortable solution to the transportation needs of the pilgrims, a feasibility study was carried out wherein the use of airships was considered.

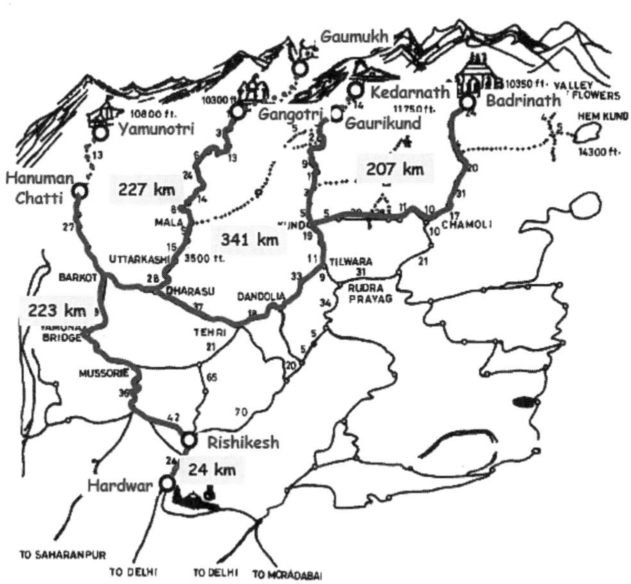

Figure 13.3 Distance by road network of various pilgrimage sites of the Char Dham Yatra

Possible Solutions

The study carried out mainly focused on comparing the conventional modes of transportation to an airship, taking into consideration the distances and the time required for travel. A comparison of the initial and operating costs were also drawn up in order to determine the best suited mode of transport for the Char Dham Yatra.

Selection of Airship

Topographical maps of the specified Himalayan region were studied and the altitudes of the desired places were recorded. Table 13.7 shows the recorded altitudes of the various points. It may be noted that none of the existing airships can operate with any meaningful payloads at these altitudes. However, some existing airships can be modified to meet these requirements.

Table 13.7 Altitudes of various pilgrim locations

Location	Altitude (AMSL) (m)
Haridwar	295
Rishikesh	356
Uttarkashi	1158
Yamunotri	3235
Gangotri	3048
Kedarnath	3581
Badrinath	3133

In a proposal for joint development of airships for operation over mountainous terrain, Airship Management Services (which is one of the most experienced airship operators in the world today) indicated that it would be technically feasible to modify a Skyship 600B airship (hereinafter referred to as Skyship 600X) to operate with 13 passengers at these altitudes [5]. The cost of such an airship is expected to be around US$6.0 million, and existence of such an airship has been assumed in this study.

Aerial distances between the various pilgrimage spots were recorded, and their comparative values with road distances are listed in Table 13.8.

Table 13.8 Comparative distances by road and air between the various pilgrimage locations

No.	From	To	By Road	By Airship
1	Haridwar	Rishikesh	24	20
2	Rishikesh	Uttarkashi	145	75
3	Uttarkashi	Yamunotri	223	31
4	Yamunotri	Gangotri	227	45
5	Gangotri	Kedarnath	341	80
6	Kedarnath	Badrinath	207	60
7	Badrinath	Haridwar	328	142
		Total	1495	453

From Table 13.8, we can conclude that the distance to be travelled by an airship is nearly one-third of that to be travelled by road. Hence, the overall time required for completing one entire cycle will decrease drastically. Considering the various other attractive features of an airship (e.g., lower infrastructure requirements), it is quite evident that airships can prove to be a better option for transporting passengers in this region.

Cost Benefit Analysis (CBA)

A cost benefit analysis was carried out to demonstrate the economic viability of the use of airships for Char Dham Yatra. A few basic assumptions were made while doing so, which have been listed below.

- A luxury car and a sports utility vehicle (SUV) are considered as alternative modes of transport to an airship.
- A total of 130 passengers are transported every day, over a seven-month period lasting from May to November.
- The entire Char Dham Yatra trip is completed in a single day if airships are considered, but requires seven days if road travel is considered.

A general layout of the cost benefit analysis for airships has been highlighted in Figure 13.4.

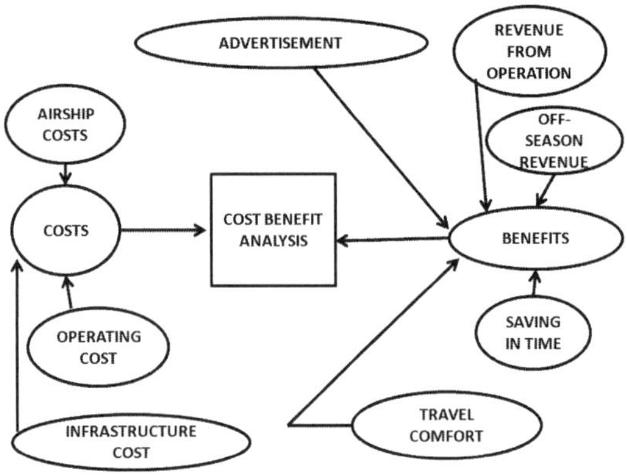

Figure 13.4 General layout of CBA for airships

Airship Operating Cost Estimation

Further to the general assumptions, the basic parameters for calculations that have been taken into consideration have been listed below in Table 13.9. These parameters have been assumed after conducting a detailed literature survey of various airships that are active in operation currently.

Table 13.9 Basic parameters pertaining to the Skyship 600X airship

No.	Parameter	Value
1	Passenger capacity	13
2	Operating speed	80 kmph
3	Fuel consumption	12 gallons per hour
4	Acquisition cost	US$6.0 million
5	Annual insurance premium	2.0% of hull cost
6	Annual maintenance and spares	6.2% of hull cost
7	Annual depreciation	10% of hull cost
8	Annual cabin and ground crew cost	US$0.4 million
9	Maximum permissible monthly utilisation	200 hr
10	Airship operational life	18,000 hr
11	Hangar cost	US$1.12 m
12	Ground support equipment cost	US$ 0.5 m
13	Revenue from operation	25% of operating cost

Luxury Car and Sports Utility Vehicle (SUV) Cost Estimation

Keeping in mind the comfort level of the passengers, it was assumed that not more than three passengers will be accommodated in a luxury car; hence a total of 44 cars will be required to operate per day. The SUV was assumed to carry seven people per trip, resulting in a requirement for 19 vehicles. The mileage for both the vehicles was assumed to be 10 km per litre of diesel fuel, costing $0.74 per litre. Table 13.10 lists a comparative analysis of the operating costs of the airship and the two types of road vehicle.

Table 13.10 Breakdown of annual operating costs (US$1,000)

No.	Item	SUV	Luxury Car	Airship
1	Vehicle cost	12.00	20.00	6000.00
2	Support equipment	0.00	0.00	2000.00
2	System acquisition	12.00	20.00	8000.00
4	Insurance premium	0.24	0.40	160.00
5	Fuel	3.00	3.00	118.78
6	Maintenance and spares	1.00	1.75	372.06
7	Crew	2.00	2.70	400.00
8	Hull and G S infrastructure	1.20	2.00	800.00
9	Annual DOC	7.44	9.85	1850.84
10	DOC per seat per trip	0.0367	0.1132	0.7119

Comparative Analysis of the Three Options

Two operating scenarios were considered while carrying out a comparative analysis, one in which the revenue was assumed to be 25 per cent higher than the operating costs and the other in which certain revenue per km of travel was assumed. The total investment required was assumed to include the cost of creating and maintaining the support infrastructure (e.g., five hangers, each to store two airships, and four masts, one at each location). The results obtained are listed in Table 13.11.

Table 13.11 Comparison of two operating scenarios for the three vehicles

Cost Item	SUV	Luxury Car	Airship
No. of vehicles required	19	44	10
Total vehicle purchase cost (US$ million)	0.23	0.88	60.0
Infrastructure cost (US$ million)	0.06	1.06	7.60
Initial investment (US$ million)	0.288	1.94	67.6
Assuming fixed revenue per km per trip			
Revenue in US$ per km per trip	0.25	0.40	25.0
Total investment (US$ million)	1.2	5.2	369.7
Years required to break even	6.5	7.6	16.3
Assuming 25% profit in operation			
Total investment (US$ million)	1.44	9.70	338.0
Years required to break even	8.2	17.9	14.6

Based on these calculations, the following observations can be made:

- Compared to the SUV and luxury car, the initial investment required in setting up an airship based transportation system is ~ 250 and ~ 35 times higher, respectively. This is mainly due to the high costs of procurement of airships, and the need for setting up a large amount of support infrastructure.
- Inclusion of value of travel time (VOTT) of passengers in calculation of generalised costs of travel can strongly tilt the argument in favour of airships. For instance, compared to a luxury car, an airship results in a net travel time saving of six days, at an additional cost per person per trip of around US$600. Thus, if a VOTT of > US$100 per day is considered, airships turn out to be a more attractive option. Further, travel by airships can definitely be expected to be much more comfortable, since road journeys over mountainous terrain can be very tiring.
- Due to their large envelope size, airships offer an additional opportunity for revenue generation, viz., advertisement and product promotion. The revenue earned through these sources can be quite high, and may even exceed the amount being charged to the passengers!
- The break-even point is a strong function of the revenue model assumed. Although the SUV needs the least amount of time for breaking even, the revenues generated are also quite low.

A comparison of investment required and revenue earned for SUVs, luxury cars and airships over several years is shown in Figures 13.5–7, respectively.

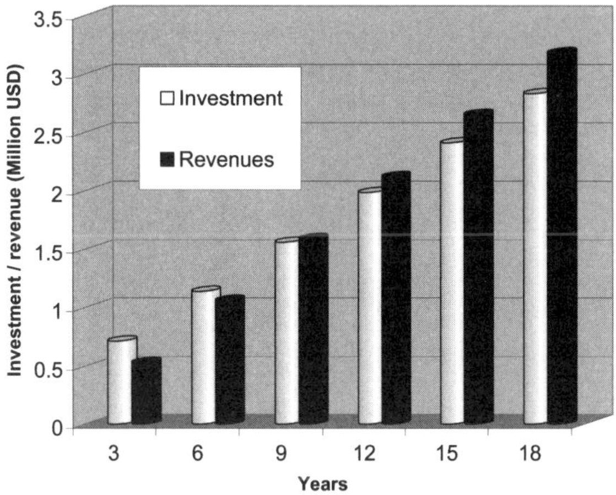

Figure 13.5 Break-even point calculation for SUVs

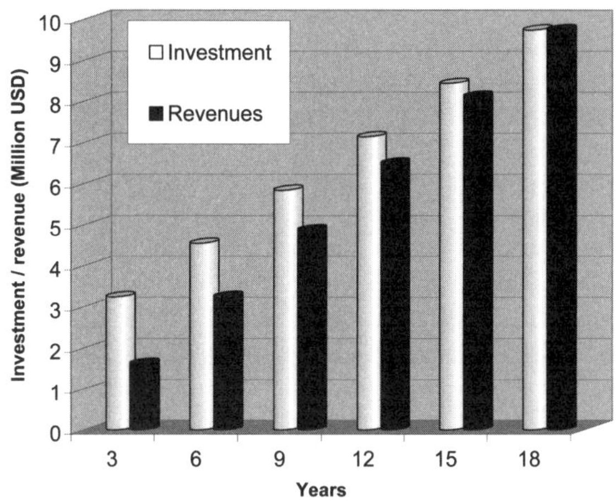

Figure 13.6 Break-even point calculation for luxury cars

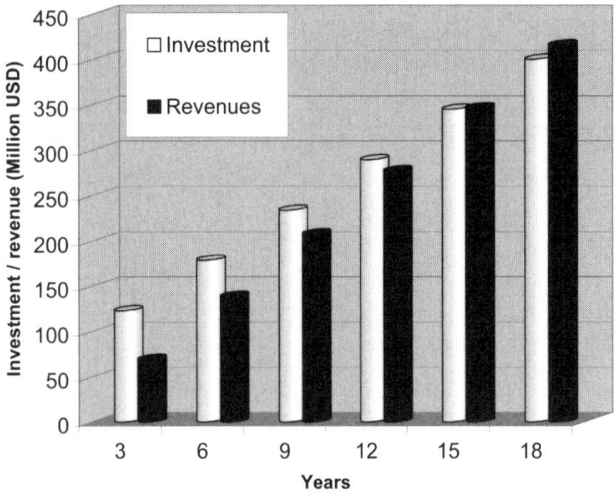

Figure 13.7 Break-even point calculation for airships

Future Pointers

Recent studies, e.g. [14], indicate that nearly 60 per cent of all flight operations today are conducted by regional airlines which mostly fly over short stage lengths. Airship experts from across the world now believe that airship passenger travel will be more competitive with jet travel as the overall trip distance decreases. This is because as the distance decreases, the total journey time is mostly a function of the access time to destination at either end of the journey rather than the flight time. The point-to-point travel time can be greatly decreased if the vertical take-off and landing capability of airships is harnessed. This reduction in time will be enough to make the airship quite acceptable for short-distance air travel. In addition, ever increasing fuel prices will further aid the development and realisation of airships in the near future.

The modern world is served by centuries of infrastructure investment in the current air, land and sea modes of transport. One of the most important factors that concerns environmentalists is the effect of greenhouse gas (GHG) emissions from aerial vehicles, mainly planes. This problem can be handled by providing improved road and rail transport connectivity and eventually reducing the use of aerial modes of transport for short distances, but the cost of investment for the same is very high compared to other alternative modes of transport.

Many countries across the world have land mass with no permanent roads. Northern resource opportunities in mining and energy across these countries are known, but are unattainable because of infrastructure cost and environmental concerns. In order to address these needs, airships prove to be the best possible alternative available. With an increase in scale of new generation airships, they can also play a role in serving intercontinental air freight markets.

Like many green technologies, commercial airships could have spin-off benefits that supplement or exceed the direct payback to their developers. The full benefits of airship innovation will be shared with users in urban centres as well as remote and undeveloped areas of the world. Soaring fuel costs and increasing concern about aviation's contribution of GHGs to the environment is now directing investors to reconsider the potential of airships.

Aviation enthusiasts and researchers are slowly beginning to believe that airships are a part of the solution set for dealing with global warming, in addition to providing an affordable and sustainable air transportation option to the developed and developing countries. With growing transportation needs across the globe, it is time that the "sleeping giant" is awakened to reclaim its rightful place in the world of civil aviation.

References

1. Anderson, J. D., "*Fundamentals of Aerodynamics*", 5th edition, New York/ London: McGraw-Hill, 2010.
2. Khoury, G. A. and Gillett, J. D., "*Airship Technology*", Cambridge: Cambridge University Press, 1999.
3. Burgess, C. P., "*Airship Design*", New York: The Ronald Press Company, 1927.
4. Liao, L. and Pasternak, I., "A review of airship structural research and development", *Progress in Aerospace Sciences*, 45, 83–96, 2009.
5. "Program on Airship Design and Development", http://www.aero.iitb.ac.in/ ~airships/
6. Gazder, R. P., "*Feasibility Study of Airships as a Viable Means of Transportation in Uttaranchal*", Undergraduate dissertation, Department of Aerospace Engineering, IIT Bombay, October 2002.
7. Jayan, A., Shankar, P., Bhaskar, K., Pramod, A., Henry, D. and Jayakrishnan, V. E., "*A Case Study on Airship Operation for Tourism in Uttaranchal*" Project Report, Program on Airship Design and Development, Aerospace Engineering Department, IIT Bombay, 2001.
8. Anonymous, "*The Model 138S Airship Specifications*", US-LTA Corp., Oregon, 2002.
9. Jackson, P. (ed.), "*Jane's All the World's Aircraft*", Jane's Information Group, 2000.
10. Anonymous, "*Bell 407 & 206L-4 Product Specifications*", Bell Helicopter Inc, USA, 2006.
11. "*Demographic Breakup of Uttaranchal*" National Atlas & Thematic Mapping Organisation, Department of Science and Technology, Government of India, 2001.
12. Dexter, J., "*Airship Operations in Uttaranchal – Feasibility Study*", Director of Flight Operations, The Lightship Group, USA, 2001.

13. *Char Dham Yatra*, Wikipedia Foundation, Inc., http://en.wikipedia.org/wiki/Char_Dham.
14. Babikian, R., Lukachko, S. P. and Waitz, I. A., "Historical Fuel Efficiency Characteristics of Regional Aircraft from Technological, Operational, and Cost Perspectives," *Journal of Air Transport Management*, 8/6, 389–400, 2002.

Chapter 14
The ICAO's Tourism Development Route Scheme

Ian Lowden and David J. Bentley

Introduction

This chapter looks at a project undertaken by the International Civil Aviation Organization (ICAO) to develop air services to peripheral and under-developed regions. The project has been in discussion for nearly three years but has not yet been actioned. We look at the process and benefits if this programme and examine closely the forerunning route development funds that have been widely adopted in Europe and elsewhere. The Essential Service and Tourism Development Route scheme (ESTDR) has not been adopted and it may be a matter of time before it is superseded by some other, perhaps less ambitious programmes.

Background to ESTDR

The Essential Service and Tourism Development Route scheme (ESTDR) is the result of a study conducted by the World Trade Organization (WTO) and the International Civil Aviation Organization (ICAO) in 2005.[1] Primarily, the study concentrates on the development potential of non-viable routes linking the tourism-generating countries with those that are less developed.

The ESTDR is recognised by the WTO as a vital element of the Sustainable Tourism – Eliminating Poverty (ST-EP) programme. It also represents the area of interest for such institutions such as the European Commission (EC), the New Partnership for Africa's Development (NEPAD), the United Nations Conference on Trade and Development (UNCTAD) and the World Economic Forum.[2]

The scheme is based on two main concepts; the Essential Air Service (EAS) scheme to develop passenger air service, together with the Essential Tourism Development Route concept (ETDR), which focuses on the international routes between specified tourism-generating markets and tourism-receiving markets. According to the ICAO study, states considered to be in most need of such schemes are a group of 50 classed as Least Developing Countries (LCDs) and Small Island

1 ICAO/WTO (2005), *A Study of an Essential Service and Tourism Development Route Scheme*, http://www.icao.int.

2 http://www.dailynews.lk/2005/07/13/fea01.htm.

Developing States (SIDS) or Land-Locked Developing Countries (LLDCs), of which African states make up a significant proportion.

The Objectives of ESTDR

Two inter-related objectives have been established for ESTDR, with a major emphasis being put on the second one. Firstly, to assure 'lifeline' provision of air services for remote or peripheral destinations and secondly to facilitate and drive economic development and develop tourism – a scheme that was developed in conjunction with the ICAO/WTO.

The first objective applies in particular to the situation of transition to liberalisation, but also in a market that is already liberalised. Moreover, it is related to other social and political elements, including keeping international links to major destinations (especially for SIDS and LLDCs) and maintaining cultural contacts with concentrations of their overseas populations.[3] The idea is that the state can support air services in order to assure maximisation of the economic and social benefits.

The second objective incorporates stimulation of inward tourism and investment, in order to alleviate the poverty of the LCDs and to stimulate their development. Tourism is being recognised here as a vital instrument in job opportunities creation and welfare development, as well as economy diversification. It concerns the most remote areas, which attract tourists for their cultural, wildlife and landscape values.

In order for LCDs to tie this development policy into their broad poverty reduction plans, it is recognised as being important to identify tourism as the major purpose of their Millennium Development Goal (MDG), Poverty Reduction Strategy Paper (PRSP) or Country Assistance Strategy (CAS) achievement plans.

Types of Intervention That Exist Currently in the UK

Direct financial support for the airline industry is currently widespread and involves, at the least, granted subsidies involved at each state's airline industry's establishment phase. However, the increasingly open and competitive marketplace currently makes it extremely difficult to offer support compatible with the liberalised air transport environment.

Air Route Development Funds (RDFs) currently come in many forms: airport support – shown as discounted landing fees (costs absorbed by airports and funding granted to allow airports to give deeper discounts); marketing and partnership funding; infrastructure investment that is less targeted but not classed as state aid

3 http://www.dailynews.lk/2005/07/13/fea01.htm.

if in a development region; and public service obligation (PSO) protecting routes from competition to allow essential lifelines to become viable.There are also a wide range of types of support including other less obvious forms such as access to slots, and training.

Some of the most publicised schemes are the funds that were established or considered in Scotland, Northern Ireland, Wales, the North West of England and the North East of England.[4] The major programmes are described below.

Air Route Development Fund for Northern Ireland

The province of Northern Ireland has historically been poorly served with direct scheduled international air services. The devolved administration identified route development as a key component of industrial and (inward) tourism competitiveness. Following Scotland's lead (see overleaf) it established a Fund in November 2003, worth GB£4 million, spread over three years, with a possible extension. Prior to the launch of the Fund, Northern Ireland had a single daily international service from Belfast to Amsterdam.[5]

Since the Fund was instigated, over 20 new scheduled routes have been launched, of which nine routes received direct investment from it. (In other words the existence of the Fund alone can be a stimulator to airlines to commence a route even if not supported, by way of a halo effect,) A major achievement was the addition of a new daily Belfast–New York service that was launched in summer 2005, largely as a result of the Fund (Figure 14.1).

The Fund sought to encourage the development of new routes by providing investment support to airports. The airports offer discounts on aeronautical charges to airlines introducing new routes, and the Fund then provides matched funding which allows the airports to double the amount of discount offered. The Fund will only do this if an independent economic appraisal of the route proposal shows that the amount invested will result in a net economic benefit to the province of Northern Ireland.[6]

---

4 Technically, RDFs in Northern Ireland and Scotland are not considered to be UK Route Development Funds, a term that applies specifically to those in, or projected for, North West and North East England, and Wales.

5 Dublin Airport, 100 miles to the south in the Republic of Ireland, exercises a strong pull on Northern Ireland air passengers. Thirty three per cent of the population of the whole of Ireland lives in Northern Ireland or the border counties but far fewer fly from the two Belfast airports or Londonderry, preferring Dublin where there is a wide range of low-cost flights. Surface links between the two cities have been improved by completion of a new motorway, and the railway line is no longer bombed.

6 The Fund's investment is limited to the first three years of operation of the new routes. Beyond that period it is anticipated that the routes will be self-supporting. Investment, which is paid on a per-passenger basis, is also limited to load factors of up to 75 per cent.

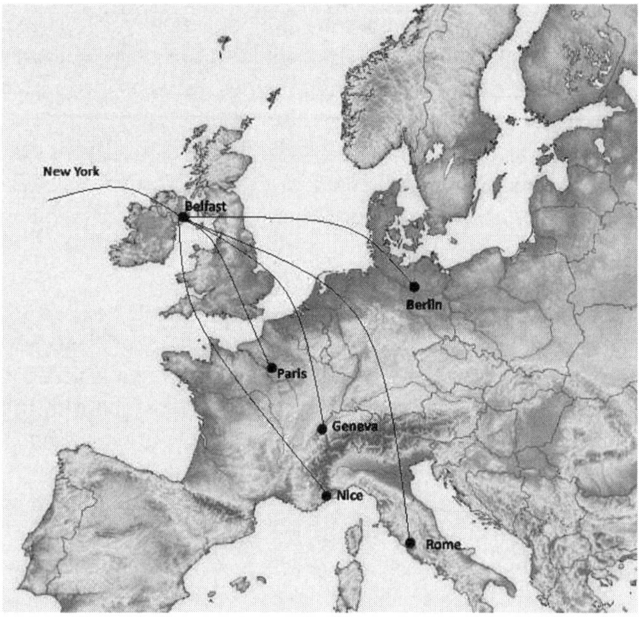

Figure 14.1 Major routes supported by the RDF in Northern Ireland

The Fund's duration has come to an end and can be regarded as a success, especially in the arena of LCC route development, where much greater penetration was needed to compete with Dublin Airport. Traffic increases have levelled out at Belfast International, and traffic actually decreased at Belfast City in 2006; but City of Derry Airport, the one most isolated by Dublin's magnetism, increased its passengers by 70 per cent in the period. The economic impact of these routes on job creation, and wider economic benefit, is currently being appraised.

Air Route Development Fund for Scotland

The Scottish Route Development Fund is a form of a public-private partnership (PPP) between local governments and private businesses. It was established in November 2002 in a partnership involving the Scottish Government, Scottish Enterprise, Highlands and Islands Enterprise and Visit Scotland to improve business connectivity and inbound tourism access all year round.

Scotland is now a semi-autonomous part of the UK with its own revenue raising and dispersal capability. The lack of key transport links to the US and some European destinations was seen as a barrier to economic growth in Scotland. The Fund's initial budget was GB£6.8 million, spread over three years, and subsequently increased to GB£13 million. It was estimated that the total economic benefit, over ten years, would be worth more than GB£300 million and that it would create around 700 tourism jobs.

The RDF promotes routes, mainly those to Continental Europe, the US and the Middle East, by supporting investment and allowing lower landing charges for the development of new routes.[7]

Scotland has traditionally seen most of its traffic 'funnelled' through the major hubs of Heathrow, Amsterdam, Paris and Frankfurt. Thanks to the RDF this has changed, creating better opportunities for the business and leisure community. By May 2008 there were nearly 40 routes operating largely thanks to the RDF.

In the 14 years between January 1989 and January 2003, the total number of scheduled international destinations served from Scotland on a year-round basis remained constant, at 17. Following the inception of the Fund, a further 15 new international scheduled destinations were quickly added (Figure 14.2). The Fund has brought in prestigious full-service long-haul airlines such as Continental Airlines at Edinburgh Airport and Emirates at Glasgow International Airport, but it also helped set up a variety of low-cost carrier (LCC) routes, including those from Scotland's main low-cost airport, Glasgow Prestwick, all by Ryanair, an airline that was already *in situ* there. Other LCCs whose start-up routes benefited are easyJet, Flyglobespan, Germanwings, Centralwings, SkyEurope and Wizz Air.

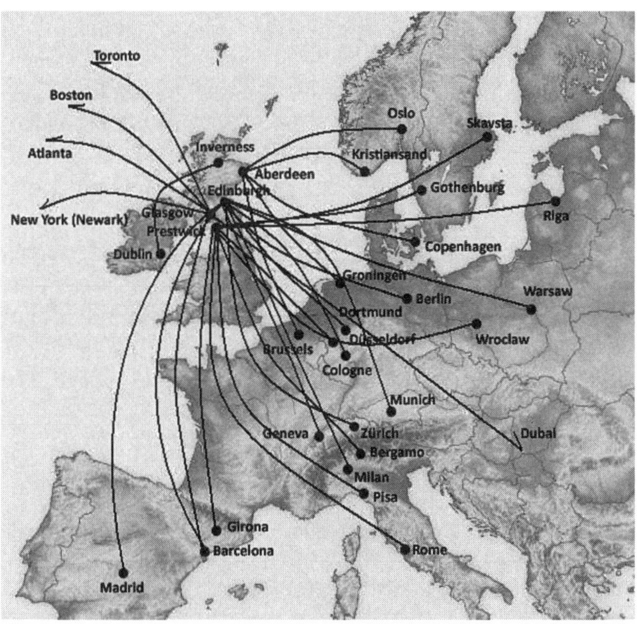

Figure 14.2 RDF international year-round services in northern Scotland, summer 2007

7 http://www.icao.int/icao/en/atb/epm/Ecp/EssentialServicesStudy_en.pdf.

There was little immediate benefit to Aberdeen Airport, which supports the North Sea oil and gas business, a labour-intensive operation that attracts manual workers from large metropolitan conurbations and which could benefit from LCC operations. Several new routes did start there, but all by full-service airlines. More recently the airport began to attract budget airlines and now hosts the main British LCCs, Ryanair and easyJet, plus bmibaby and local airline Flyglobespan. It was the fastest growing BAA Scottish airport in 2006 and 2007.[8] Strict criteria needed to be met before RDF funding was allocated to Scottish airports. Essentially, a new external direct route must be of economic benefit to Scotland and predominantly of long-term benefit to business travellers, although it also allows for strong inbound tourism routes. It must operate at least five days a week, all year round and not compete with an existing service. A full economic appraisal is carried out for each prospective route, scored on factors including destination, creation of jobs, number of inbound tourists and journey time saving.

The funding of new routes by the RDF stopped in May 2007, although the routes that had started continued to receive funding if they were eligible. In view of the success of the scheme the Scottish Government began looking at alternative ways forward but no clear process has been decided upon.[9]

Air Route Development Fund for Wales

Another recent UK RDF was in the Principality of Wales, which is heavily populated in the south in the Cardiff (capital)–Swansea belt but thinly populated in other areas. The Welsh Assembly introduced an Air Route Development Fund in the (northern) summer of 2006. The Transport (Wales) Act 2006 provides the Assembly with the powers to give financial assistance for air transport services in Wales.

This fund, as with the others discussed above, focused on 'sharing risk' while routes matured and became viable. Again, the focus was on a reduction on aeronautical charges and marketing support of up to 50 per cent of the total costs. This has led to a number of successful start-ups from Cardiff.

The RDF Protocol ended on 31 May 2007. The European Commission[10] published its guidelines on financing of start-up aid to airlines departing from

8 CAA statistics.

9 http://www.scotland.gov.uk/Topics/Transport/19675/air-route-dev-fund/about-air-rdf.

10 In the event that funding is only available from public sources, RDFs would have to comply with EC Guidelines. Key principles were that much of the decision relates to specific circumstances of the relationship between the Walloon Government (Belgium), Brussels South Charleroi Airport and Ryanair following a legal challenge on perceived state aid to Ryanair at Charleroi Airport by other airlines. The Guidelines confirm that regions and airports can support the introduction of new air services, subject to the key principles being followed. The critical test is the 'market investor principle' but beyond this support can

regional airports in December 2005, with one of the main changes proposed being the reduction in aid that could be offered from 50 per cent to 30 per cent from 01 June 2007. The conditions and restrictions contained in the new RDF Protocol made the scheme unattractive because of the disproportionate work required to obtain a 30 per cent contribution. As a result, the Welsh Route Development Fund closed to new services that commenced after 31 May 2007.

Route Development Schemes Outside the UK

Outside the UK there has been less of a commitment to overtly fund new routes. Most prominent have been the European PSO schemes in France, Greece, Ireland, Portugal, Spain and also in the UK.

Europe: The Public Service Obligation (PSO) Scheme

The EU Second Package of Air Transport Liberalisation (1990) and the Third Package (1993) created the platform for the Public Service Obligation (PSO) scheme. This allowed member states to support domestic and intra-EU international routes to protect and develop scheduled air services to peripheral, development region or thin routes at regional airports. The aim is to support routes that are essential for economic development until they are profitable. There are now around 150 PSO routes having either market protection, subsidies or both.

Member states that designate PSO routes then issue a public invitation to tender for each route (or a group of routes in the case of Norway). The invitation is open to all airlines registered in the European Economic Area. It sets out minimum service levels in terms of capacity, frequency and schedule and any limits on the level of fares, or the need to offer specific fare types. There are often timetabling requirements – some to allow day-return trips and others to allow convenient onward connections – and there may be conditions related to the type of aircraft that can be employed (with pressurisation, for example). Initial tendering is on the basis of no subsidy, but with regulatory protection from competition. If no carrier is willing to meet the requirements without subsidy, the invitation is reissued, this time offering subsidy, usually for a three-year period. The selection of the carrier must take into consideration the adequacy of the service, the fares and the level of subsidy required. Contracts usually last for three years, after which they are reviewed. The greatest numbers of PSO routes in 2004 were within Norway (61) and France (49), with the then total for the whole of the EU of about 180.

include airport charges, discounts and contributions to start-up costs (including marketing) and it should be for services to destinations not presently served from the airport. Support must be open to all airlines wishing to start services under similar circumstances (to the same destination and with similar characteristics in terms of frequency, type of destination etc.).

On French, Italian and Portuguese PSO routes linking islands to the mainland, operators are required to provide discounted fares of up to 40 per cent to island residents, students and senior citizens.

PSOs can be administered, awarded and funded by both national and regional governments, either directly or indirectly through associated agencies. Almost all PSO routes are domestic, although there is no rule that precludes an application for an international service.

USA: The Essential Air Service (EAS) and Small Community Schemes

These were designed to subsidise services between remote communities and hubs and they are enhanced by 'travel banks' funded by corporations and marketing funds, typically through visitor and convention bureaux.

The EAS programme was established in 1978 to ensure that no communities would lose air services as a result of the Airline Deregulation Act of that year. The Department of Transportation (DoT) determines both the eligible communities (e.g. those over 70 driving miles from a large or medium hub airport and where the subsidy required does not exceed US$200 per passenger) and the required service levels (a connecting hub airport, frequency, capacity etc.) for each community.

A scheme for ensuring services to small communities began in 1958 based on a mail rate subsidy and this was adapted to form the current EAS programme. More than 100 communities in the contiguous states and Hawaii and 35 in Alaska benefit from the programme. The choice of carrier to benefit under the system is not based solely on the lowest bid – other factors considered include the carrier's experience and the arrangements for through ticketing and code sharing. The programme allows the DoT to take steps to ensure that slots at congested airports are made available for EAS services.

The level of support often results in two to four round trips per day to a major hub. Awards are usually made on a two-year basis, with operators entitled to reapply for the franchise. Payments are made monthly in arrears on an 'after the fact' basis. Operators wishing to terminate, suspend or replace a service have to give 90 days' advance notice. The average population size of the communities served is about 120,000 people – although when the whole catchment area is taken into account this increases to 175,000. The annual cost of the programme is about US$100–120 million, with grants for individual routes ranging from US$0.29million per year to US$1.87 million per year.

In addition to the EAS programme, the Small Community Air Service Development Programme was introduced in 2000 to run alongside it. This programme has granted a total of about US$20 million to a maximum of 40 communities served by an airport that is not larger than a small hub airport and with insufficient air services or unreasonably high air fares. Grant funds can be used, for example, for financial incentives (including subsidies and revenue guarantees) to airlines and to cover the expense of new promotional activities related to improving air services (ICAO/WTO 2005).

It is designed to help smaller communities (or consortia of communities) enhance their air services and address issues related to high air fares – although communities with no existing services are able to apply and some grants have been given to cover the costs of feasibility studies. Priorities for an award are where: the average air fares are higher than normal; a portion of the costs of the planned activity is provided from local, non-airport revenue sources; a public-private partnership will be established to facilitate air carrier service; and the funds will be used in a timely fashion. A number of applications are received from, and awards have been made to, communities already benefiting from subsidised services under the EAS that are seeking to promote those services or expand their service provision. Grants can be used to provide assistance with marketing, hiring additional staff or for service improvements, including increasing the number of airlines on a route, increasing frequency and upgrading the service to use larger aircraft.

The US DoT is authorised to award a maximum of 40 grants per year, with no more than four in any one state. By mid-2005, a total of 68 communities had benefited from the programme. The awards are limited to three years. The annual cost was about US$20 million in 2001, rising to US$35 million for each of the years 2004–08. Three-year grants in 2005 ranged in value from US$85,000 to US$1.6 million each. A unique feature of the programme is that considerable flexibility exists in terms of what the grants can be used for.

The Developing Countries

In the developing world there are Caribbean region subsidies – revenue guarantees and subsidised fuel for airports – and in Africa there are the Egypt, Tunisia and Morocco programmes of route support for charter flights. Route support schemes here are all 'bespoke' and lack consistency. The member states of the West African Economic and Monetary Union (WAEMU) adopted a liberalisation package of the common air transport programme within the area in 2002, including a PSO scheme similar to that of the EU. (This is not to be confused with the wider, Pan-African Yamoussoukro Agreement that remains to be implemented fully.) Several small island states and dependencies in the Caribbean and Pacific whose economies are dependent on tourism have provided financial assistance to airlines such as direct subsidies for the operation of a route and the purchase of a specified number of seats to keep their traffic linked to tourism generating developed economies. Member states of the Pacific Islands Forum have been discussing the mechanisms to support essential international air services under the Pacific Islands Air Services Agreement. Separately, the Solomon Islands investigated a scheme of its own in 2006–07.

Australia: The Remote Air Service Subsidy Scheme

The objective of the Australian Remote Air Service Subsidy Scheme (RASS) is to provide communities in remote and isolated areas with improved access through a regular weekly air transport service for the carriage of passengers and goods (e.g. educational materials, medicines, fresh foods and other urgent supplies). Because of the distances involved and the fact that road access to many of these communities is often cut for several months during the wet season, a regular air service might offer the only reliable means of transport. Around 30 per cent of points served are traditional indigenous communities, while others include cattle and sheep stations.

Under the Government of Australia's scheme the airlines win support through a competitive tender. The term of the support is fixed under an agreement with the Australian Government and the subsidy is paid directly to the airline. The airline must offer at least a weekly scheduled service; several of the operators provide a RASS-based service for Australia Post to ensure delivery of mail and packages to outlying regions.[11]

To gain admission to the RASS scheme, communities must meet specific eligibility criteria. The capacity to admit communities is limited by the scheme's budget in any single year. The level of subsidy is relatively small and the total budget was less than AU$3.6 million (in 2005–06) but this has been used to benefit 225 communities in remote parts of Queensland, Northern Territory, Western Australia, South Australia and Tasmania. The value of the awards made to individual airlines is made public.

The community has to make a considerable effort to gain the subsidised air service and it must demonstrate that there is demand for a weekly scheduled service, but that it is also sufficiently remote from another city or community having a weekly scheduled service for it to be justified.

Australia also has a scheme to help small operators pay the air traffic control charges of AirServices Australia. This scheme, which commenced in January 2002, is provided to assist regular public transport and aero-medical services operators with the payment of en-route air traffic control charges. All airlines that operate regular public transport (RPT) services using aircraft with a maximum take-off weight of 15 tonnes or less are eligible for assistance under the scheme. Operators of aero-medical services, such as the Royal Flying Doctor Service, are also eligible.

Airlines that operate solely within Western Australia and service single operator routes are eligible to register RPT aircraft that have a maximum take-off weight of up to 21 tonnes (this corresponds approximately to a 50-seat aircraft). The Australian Government has announced that it will review the effect of Airservices Australia's charges on regional airlines to ensure continued growth of the sector is not being impeded.

11 http://www.infrastructure.gov.au/regional/indigenouscomm/rass.aspx.

The Principles of ESTDR

Building on the global experience of supporting air services and the needs of the developing peripheral communities, the ICAO and the United Nations World Tourism Organization (UNWTO-OMT) have developed a conceptual framework for a scheme that would help develop air service through the support of essential air routes. ESTDR is identified by the ICAO (ICAO/WTO 2005) as:

> a mechanism whereby support, in the form of a financial subsidy and/or an exclusive concession, can be provided to airlines for the provision of certain services of a public service nature.

The scheme encourages tourism in poor countries becoming, therefore, a part of their recovery plan. The support is usually in the form of generic direct subsidy, e.g. the part funding of fuel provision, en-route navigation or airport charges. There may also be some indirect subsidy, e.g. the part funding of marketing, heavy maintenance, administration or loan guarantees or an implicit subsidy, e.g. allowing and facilitating the cross-subsidisation of loss-making routes from profitable routes.

The experience of EU member states is that targeted assistance has been extremely beneficial in allowing critical 'lifeline' routes to remain viable, thereby protecting the interests of some of Europe's peripheral communities. In line with the ESTDR principle objectives, the entire mechanism and process of the scheme is transparent, accountable, non-discriminatory, sufficiently flexible and, as far as possible, market-oriented to minimise market distortion and enhance airlines' efficiency incentives.

ESTDR Criteria

The study on ESTDR by the ICAO recognises several criteria that are vital for the ESTDR design process. The principle condition, analysed further, is known as a 'two key' test. Other conditions include service level specification, carrier selection, contract duration, subsidy payments and sources of financing.

The Two Key Test

The work produced two key tests of eligibility relating to 'socio-economic objectives' and 'economic justification' and we have described these below.

Test 1: Demonstration of need – does the community need air service? The key tests are based on evidence of local need, specifically, to prove that air service is needed because the region and/or airport is very remote or that vital economic sectors cannot remain viable without air service; often the key sector will be tourism. There may also be social or political reasons such as education, access to medical centres or ethnic and cultural links.

The first test examines whether there is a demonstrated need for an adequate level of international air service, for example, routes serving 'remote or peripheral areas' and/or 'development areas' for tourism. If the airport or region meets one or some of these criteria, it is anticipated that the process then moves to the next test and then to the implementation steps described below.

Test 2: Economic justification The second stage of the test would be an economic assessment based on demand for air services and airlines operating cost conditions. Route economics will be scrutinised to understand if the air service is not immediately viable, but with the right level of support it could be viable.

Typically, the types of region attracting this kind of support would be characterised by small population, a remote location and low income levels. Furthermore, the development of air service may be hampered by poor airport infrastructure.

Structuring the Schemes

Once the key tests have been passed, the emphasis moves to structuring the scheme. Passing the test and structuring the scheme have a degree of iteration. In particular, structuring the schemes will typically entail eight stages:

1. route selection – including demonstration of need
2. service level specification
3. carrier selection
4. contract duration
5. subsidy payments
6. sources of financing
7. supplementary options
8. regulatory arrangements.

Each of these areas requires key skills or techniques such as an understanding of route economics.

Route selection The scheme is deliberately flexible and non-prescriptive, and it is intended that the participating government can use it as a structure to prioritise the most important routes. Typically, governments were expected to look at 'need' in terms of 'lifeline provision' for economic and ethnic reasons, cultural contacts and tourism. However, the scheme's authors felt that economic development should carry the most weight.

Service level specification The ICAO/UNWTO study emphasised a flexible approach in setting service levels. This phase involves determining an adequate level and minimum standard of air service (usually in terms of frequency, capacity etc.) on each selected route. A flexible and market-oriented approach is advocated so that the airlines' ability to be entrepreneurial will not be constrained (ICAO/WTO 2005).

There is a 'chicken and egg' nature to the service offering of airlines (frequency and capacity) and the passengers carried, route income and the costs of operation. Service levels are expected to be determined after a study of the likely market size, itself a major exercise requiring the study of business and leisure demand, tourism development and other 'stimulated' traffic created by the existence of the air service. There would be a need to optimise service levels against the cost of operation.

Carrier selection At this stage it is necessary to design the appropriate institutional mechanism as a means to allocate a targeted subsidy so that market distortion and unfair competition are minimised.

The process of selecting the airline will have a number of determinant factors. One immediate hurdle is the eligibility of the airline to appear on one of the countries' air service agreement. In addition, it was recommended that the lowest tender should be a strong criterion. As with the PSO discussed earlier, there may be a period of protection on the route for the new-start carrier.

In order to ensure that cost-competitive and eligible airlines are encouraged to apply there would normally be an 'auction' or bidding period publicised through the industry's main channels such as Airline Business or Routesonline.

Precise tender criteria The precise criteria will be governed by the needs of the LDC and the scale and desirability of competitive behaviour between airlines. these criteria include:

- specification of routes and/or groups of routes for which support/market protection is required
- level of support required
- market protection required
- extent of tourism development produced
- airline financial viability
- the business plan
- code sharing with partner airlines
- whether 'solus status' is required
- other, ad hoc, factors.

There will be a need to weigh these factors so that the competing bids can be compared rationally.

Duration of the contract Since the contract duration has implications for airlines' efficiency and incentives, it is determined carefully on a case-by-case basis. For long-term contracts, a mid-term review and/or enforcement, monitoring, audit and renegotiation may be necessary to deal with unforeseen changes (ICAO/WTO 2005).

Two broad options and a sliding scale between them are recognised: short-term contract with the airline – has the advantage of allowing competition onto the route or eliminating subsidies, but discourages investment, and; a long-term contract – has the advantage of allowing the airline time to develop the route and make the necessary investment.

A challenge exists here to develop a framework for changing economic and market conditions and airline management and operating performance. Hence, the trade-off is between not providing a sufficiently attractive proposition (in the case of the short-term contract) and creating a long-term protected monopoly, which would be bad for the consumer and possibly for overall market development.

Subsidy payments One of the major concerns of the process is that the airlines will be overpaid or will under-deliver. Clearly the scale of support has to be supported by rigorous auditing, and the tender process will itself regulate the airline's behaviour by being clear and transparent (ICAO/WTO 2005).

Subsidy payments should reflect the gap between costs and revenues reflecting direct operating costs (DOC) – directly associated with operating a route such as provision of fuel, en-route navigation and airport charges – and indirect operating costs (IOC): marketing, heavy maintenance, administration and overhead costs.

Airlines will monitor the performance of a route by recording the route DOC and an allocation of IOC and administration/overheads to the route. They will then assess the route on a fully allocated basis.

Sources of financing Probably the main challenge of the scheme, and partly the reason for the wind down of the European schemes, is the availability of funding. This scheme calls for states to find their own funding, something it may find difficult given the competing claims for vital infrastructure. Ideally, international agencies or private partners could be persuaded to fill any funding gap (ICAO/WTO 2005).

Therefore, the key considerations with regards to financing can be summarised as Who has the primary responsibility for raising financing? Is it 'competitive' with other state expenditure? If so, what is the cost-benefit analysis (CBA) procedure? Would the government need external domestic or international support? These factors require a deep understanding of the sources of international funding and finance aid.

Supplementary options Consideration may need to be given to a number of optional measures, which are supplementary to, or alternatives to, an ESTDR scheme. Supplementary measures, such as indirect subsidies or other support to airline or tourism partners, could contribute to reducing the absolute subsidy amounts and the number of subsidised routes in the mid- or long term with relatively smaller implementation costs, if properly designed. However, any supplementary measure should be governed by the same conditions and principles as those applied to an ESTDR scheme itself because of the potential for distortion of the market (ICAO/WTO 2005).

Pertinent, but not central, to the development of an ESTDR scheme are also considerations of the tourism facilities in an LDC that may not be adequate to attract tourists. Other obstacles may include insufficient funds to support an ESTDR scheme. There has to be consideration of which organisation can be a catalyst for change and there will be the need for technical expertise and/or advice in developing the market and developing an action plan and overseeing implementation. There will also be the need for a cross-check to ensure that the air service development plans continue to reflect the needs of the economic development and tourism plans (Figure 14.3).

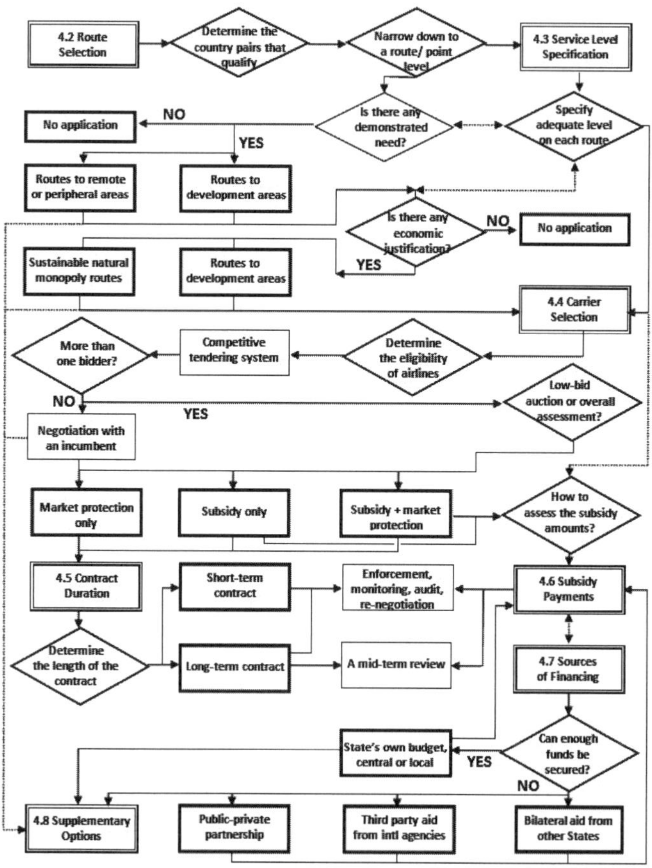

Figure 14.3 Development of ESTDR

Source: Adapted from ICAO/WTO (2005).

Regulatory arrangements ESTDR emphasises a *unilateral* approach in which the state can provide subsidies quickly. In contrast, bilateral discussions create the risk of prolonged negotiations. It is important that this is clarified as the lack of consistent and well-considered regulation can lead to disagreements between states and/or fractious complaints procedures.

The Success and Future of ESTDR

Since the launch of ESTDR there has been considerable interest in the programme, evidence of which was clearly manifest from the attendance at ESTDR seminars at Regional Routes Africa events,[12] where its potential and development were discussed. There is a common view that the success of the RDF, PSO and other route development schemes could be replicated across the LDCs to a greater extent.

The feedback from the events was that there was a need to assist and incubate air service to the periphery of LDC countries. Furthermore, it was recognised that tourism is a potentially strongly positive economic sector to develop, provided it was sensitive to social and environmental considerations. It has been generally acknowledged that air service and tourism development are interlinked, but that the former is the harder to develop because of the very high fixed costs. There is a view that air links could be the 'seeds' from which a prosperous tourism sector could be developed.

There has also been a general recognition that the ESTDR scheme was, in principal, rigorous and well constructed. This is largely because it has been developed with the knowledge of the existing air service development programmes in Europe and beyond. The ESTDR offered a common and consistent platform for the application of funding to achieve well-defiend goals.

The LDCs have noted the success in particular of the UK schemes described above; low-cost carriers are becoming increasingly common across the developing world and they have recognised the benefit they can bring. It is probably only a matter of time before they start requesting assistance beyond the normal honeymoons on airport charges. At this point the LDCs would benefit from a programme that brings control and rigour to the process. Tour operators from Europe are also increasingly mindful of the benefits they can bring and are assessing incentive schemes.

The major obstacle is clearly the availability of funding. ESTDR is a process that the ICAO-WTO advises but does not support materially. Once a programme is agreed it would be the responsibility of the sponsor governement to find the funding. This has presented a great difficulty because of the many competing claims for government funding within the LDCs. This is entirely understandable and it would be an extremely brave government that pushed the support of an

12 Regional Routes Africa is the African-based Routes meeting where airports and airlines meet to discuss route development. http://www.routesonline.com.

airline, probabaly foreign owned, over the cause of hospitals or schools, even if an economic impact study proved that it would be a wealth-generating project.

The answer may be to engage and win the enthusiasm of the world development community; ideal partners may be the World Bank or Arab Investment Bank. However, given the current uncertainty regarding aviation and its environmental impact this may be difficult.

The conclusion may be that individual countries use the best aspects of ESTDR to promote their own development programmes, leaving ESTDR as a useful and interesting platform of ideas but nothing more. Only time will tell.

PART E
Key Future Issues

Chapter 15

Aviation and Climate Change – Facing the Challenge

Olav Mosvold Larsen and Jon Sjølander

Introduction

In its fourth major report published in 2007 the Intergovernmental Panel on Climate Change (IPCC) concluded that it is highly likely that human emissions of greenhouse gases have been the main cause of the increase in global temperature since the mid-20th century. If emissions are not reduced, the global average temperature will continue to rise. Resolving the climate problem is a major social challenge and hence it is impossible to sidestep the environmental challenges when discussing aviation policy.

However, attention is often focused on how important aviation is to society. The desired development makes a growth in air traffic a necessity, especially in remoter regions. In most countries increasing tourist traffic is a political goal. Taking Norway as an example, it is also a political goal to maintain a decentralised structure of trade and industry. The public expects nationwide access to good health services, and the petroleum industry on the continental shelf is dependent on air transport. Hence, the aviation industry's objectives may seem to be pulling in different directions, and balancing the various considerations relative to one another is a challenge.

This chapter will address the climatic effects of aviation and relevant abatement initiatives for the industry. Firstly, we give a short introduction to a project initiated by the Norwegian aviation industry to respond to the climate challenge. Secondly, we will discuss the climate effects of aviation based on the findings in the project "Aviation in Norway: Sustainability and Social Benefits" (Avinor 2008a, 2008b). The main section presents the most important abatement initiatives from the above-mentioned project. The chapter concludes with a discussion of the effect of the proposed abatement initiatives and the Norwegian aviation growth perspectives, illustrating that it is possible to stabilise emissions from the industry, despite the projected traffic growth.

Facing the Challenge

The aviation industry in Norway recognises the challenges of reconciling aviation growth, social and economic benefit and environmental challenges. The Norwegian aviation industry therefore started a project in spring 2007 to promote initiatives that contribute to sustainable and socially beneficial aviation in Norway. There were three major objectives for the project:

1. to investigate initiatives to reduce the negative environmental impact of the industry;
2. to give a comprehensive presentation of the environmental impact of the industry;
3. to contribute to a comprehensive presentation of the social and economic benefit of the industry.

The project was initiated and managed by Avinor, but the Norwegian aviation industry collectively – represented by Avinor, the Federation of Norwegian Aviation Industries, SAS Norway, Norwegian Air Shuttle and Widerøes Flyveselskap – was responsible for the project and the research and documentation on initiatives to reduce the industry's negative environmental impact.

To ensure a thorough and transparent process, the Norwegian Confederation of Trade Unions (LO), the Norwegian Pollution Control Authority (SFT) and the main Norwegian environmental non-governmental organisations – Bellona, Friends of the Earth Norway and The Future in Our Hands – participated in a "resource group". The group discussed topics related to aviation and sustainable development and was, through regular meetings, informed about the progress and the findings of the project. The resource group was however not responsible for any results or conclusions of the project.

The efforts to coordinate and develop abatement initiatives covered the entire travel chain. Work groups for 1) surface access to/from the airport; 2) airport operations and 3) aircraft engineering and aircraft operations were established. In addition, an expert report on the environmental impact and social and economic benefits of aviation was commissioned. The work groups held several meetings and came up with more than 50 detailed proposals to reduce greenhouse gas (GHG) emissions from the aviation industry. The initiatives will be further elaborated below.

During the last years it has become harder to process the vast amount of information available, especially on the environmental impacts, but also on the social and economic benefits of the industry. Thus the goal of the third party report was to collate this information in one volume, to serve as a reference document in the national public debate. The Norwegian Institute of Transport Economics (TØI) and Cicero – the Centre for International Climate and Environmental Research – contributed background materials to the presentation of the social and economic benefits and the environmental impacts of the industry (Lian et al. 2007a). Asplan Viak AS provided analyses and background materials to the work group on surface access to/from Bergen and Stavanger airports (Asplan Viak 2007a, 2007b).

The next section will focus on the environmental challenges. Several other findings from the project will not be disseminated in detail here.

Environmental Facts

Aviation's Impact on the Climate[1]

Aviation emits several gases and particles and thus affects climate change and air quality. Emissions from aviation change the atmosphere's composition of particles and gases, and a significant share of the emissions takes place in the upper atmosphere.

Emissions of carbon dioxide (CO_2) and water vapour (H_2O) have a direct effect on the climate. They are both greenhouse gases (GHG) and thus have a warming effect.[2] The direct effect of water vapour is small, but emission of water vapour generates contrails, which in turn may form cirrus clouds. Both contrails and cirrus clouds can make a significant contribution to global warming. Of particles and gases that quickly form particles, aviation emits sulphur dioxide (SO_2) due to the content of sulphur in the fuel (SO_2 forms sulphur particles). In addition, aviation emits soot and carbon particles (OC). Sulphur and carbon particles reflect the radiation from the sun and have a cooling effect on the climate, while soot absorbs the sun's rays and has a warming effect. Emission of particles may also have an effect on the formation of clouds. Emission of nitrogen oxides (NO_X)[3] has a climatic effect by producing ozone (O_3) or breaking down other greenhouse gases such as methane (CH_4) in the atmosphere, and is considered to have a net warming effect.

Emissions of CO_2 and water vapour are a product of the use of fossil fuels. The amount of NO_X emissions is, however, also dependent on the combustion process in the jet engine. At high temperature and pressure, nitrogen (N_2) and oxygen (O_2) react and nitrogen oxides (NO_X) are the result. In general, higher temperature and pressure in the engine will reduce the amount of fuel needed, but it will also increase the emissions of NO_X. Emissions of soot and particles are in general a result of "polluted" fuels; they are not needed for the combustion process in the aircraft engines, and can be limited by producing "cleaner" jet fuel.

The emissions from aviation thus consist of several gases and particles which have different impacts on the climate due to their "global warming potential", how long they stay in the atmosphere and at what altitude they are emitted. Furthermore air temperature, humidity and other meteorological aspects have an impact on the formation of contrails and cirrus clouds. The "global warming potential", "lifetime" and "altitude" are just samples of variables that impact the overall global warming potential of emissions from the aviation industry. There is an ongoing debate concerning these effects, particularly relating to water vapour

1 This section is based on Lian et al. (2007a).

2 However, only emissions of CO_2 (and CH_4, N_2O, PFK, HFK and SF_6) are regulated under the Kyoto Protocol.

3 NO_X is a generic term for mono-nitrogen oxides (NO and NO_2).

and NO_X. The argument is that the effect of emissions is aggravated when released at higher strata. This debate will be further elaborated below.

Methods of Allocation and Calculation

On a global basis, aviation is responsible for a little over 2 per cent of global greenhouse gas emissions. However, the emissions vary considerably in different parts of the world. Aviation is an international business, which raises challenges as to how the emissions should be allocated.

In most Organisation for Economic Co-operation and Development (OECD) countries, the amount of emissions from aviation has been depicted very differently in the media. Editorials and headlines in the main Norwegian newspapers did in 2007 quantify annual emissions from air traffic originating in Norway to be between around 1 million to 7 million tonnes of CO_2, allegedly the equivalent of around 2 or 15 per cent of total Norwegian emissions. This illustrates that the choice of allocation principles used to quantify emissions from international travel is crucial. The choice of principle is a difficult one, and it could e.g. not be solved during the Kyoto process. Several practical considerations must also be made to ensure robust data collection (Lian et al. 2007). Should the emissions be allocated to:

- the country where the fuel is sold;
- the "home country" of the operator or where the aircraft is registered;
- the country of departure or arrival of the flight;
- the country of origin or destination of the goods and passengers on board;
- the nationality of the traveller?

The list is not complete, but illustrates some of the available options. All have pros and cons, and the principle chosen will have impact on e.g. a country's ability to fulfil its Kyoto commitment. It may also turn into an interesting economic debate as emission of carbon dioxide now comes with a price in many countries.

Figure 15.1 illustrates the most used methods of calculation in the Norwegian public debate on the impact of aviation on the climate. One method is to calculate emissions from domestic traffic on the basis of fuel sold for domestic civil purposes. The principles used in this method are laid down in the Kyoto Protocol and are included in official national emissions statistics and reports on monitoring the targets of the Kyoto Protocol. This does not however provide a complete picture as emissions from international travel are not included.[4]

It is thus also common to calculate emissions based on the total aircraft fuel sold in the country – "bunker fuel". In practice, bunker fuel covers all domestic flights and international flights to their initial destination. The totals of all countries' bunker fuel figures are reported to the UN Framework Convention on

4 The same applies to international shipping.

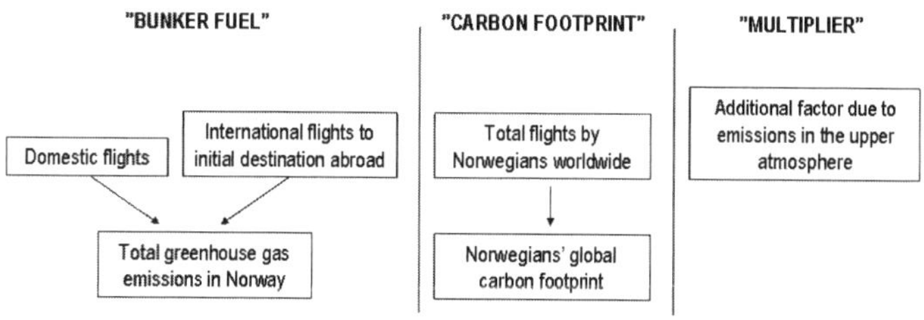

Figure 15.1 Main methods of calculating the impact of Norwegian aviation on the climate

Climate Change (UNFCCC). They form the basis of the calculations of total global greenhouse gas emissions from aviation, ensuring that all emissions are included and that there is no double accounting.

A third possible and often quoted method of calculation provides an estimate of emissions from air traffic from inhabitants of a specific country. Such calculations try to provide a picture of the inhabitants' "carbon footprint" related to journeys by air by calculating the emissions from both domestic and international air journeys, as well as those between and within third countries.

The picture gets even more complicated when moving from absolute to relative numbers. National authorities and the national aviation industry can influence domestic traffic and traffic to the first point of arrival abroad. This is also included in regular official statistics nationally and internationally, and the data collection is fairly robust. It is, however, also possible to calculate the global emissions from all air travel activities for a country's inhabitants (carbon footprint), by analysing travel-habit surveys, using emissions calculators etc. However, within the national greenhouse gas audit, it is methodically incorrect to compare total global emissions from air travel for, say, Norwegians with total domestic emissions in Norway. If such a share were to be calculated, an assessment would have to be made of Norwegians' global greenhouse gas emissions from aviation in relation to Norwegians' global greenhouse gas emissions from all consumption, export deducted and import included. There is currently no satisfactory technical or scientific basis for arriving at certain estimates for this.

Emissions in the Upper Atmosphere Require Special Attention[5]

As indicated previously, the environmental challenges of aviation are not only linked to direct emissions of CO_2. The issue has been fairly high on the agenda since the Intergovernmental Panel on Climate Change (IPCC) published "Aviation and the Global Atmosphere" in 1999 (IPCC 1999). The IPCC maintains that aviation faces special challenges because the emissions of other gases in the upper

5 This section is based on Lian et al. (2007a).

atmosphere have an added impact. It is uncertain how large this impact is and it is not regulated by the Kyoto Protocol. It is worth mentioning that activities of other industries also imply impacts on climate that are not regulated by the Kyoto Protocol. The meteorological and chemical mechanisms regarding these emissions are complex. The scientific understanding of the impact in higher strata of air is still limited. Several research initiatives have been undertaken and more are under way to shed light on the effects of these emissions.

The issue has also entered the political arena in several countries with proposals, especially from green groups, that aviation should pay higher taxes due to these emissions or that the effects should be included in environmental impact assessments of transport in general. During spring 2008, the issue was also heavily debated in connection with the directive on inclusion of aviation in the European Union's Emission Trading Scheme (EU ETS).

As mentioned above the three variables – global warming potential, lifetime and altitude – affect how the aviation industry's (and other industries') emissions impact on the climate. It is beyond the scope of this chapter to give a thorough presentation of these issues; but to illustrate the challenge we could mention that of the CO_2 emitted today, 35 per cent will still be prevalent in the atmosphere in 100 years from now, cirrus clouds have a lifetime of a few hours, ozone a couple of months and methane about ten years. Related to cirrus clouds it could be mentioned that air travel at night is worse for the climate than during the day because clouds at night prevent heat accumulated on the ground during the daytime to "escape". Furthermore, on shorter journeys, the aircraft may only be at the critical altitude for a short period of time. This is especially relevant when discussing regional air transport. When flying from Oslo to Bergen only 20 per cent of the greenhouse gas emissions will be during the critical altitude, while when flying from Oslo to Bangkok approximately 80 per cent of the emissions will take place in that altitude.

How can these effects be measured? Based on today's level of scientific understanding of these issues and a 100-year time span (as also applied in the Kyoto Protocol), it is estimated that the additional high-altitude effects could imply a multiplier of 1.2, without the effect of cirrus clouds (Forster et al. 2007), and 1.8 including cirrus clouds (Stordal et al. 2005; Fuglestvedt et al. 2008). However, the effect of contrails and cirrus clouds are uncertain, amongst other things, because of local meteorological conditions. This was one of the reasons for not applying a multiplier in the directive on aviation and EU ETS in 2008. Applying a high-altitude multiplier with higher numerical value can be interpreted as either a deliberate choice of shorter time frame, or to give weight to the precautionary principle due to scientific uncertainty about these additional effects.

The Industry's Planned Initiatives for Reducing Greenhouse Gas Emissions

Air traffic constitutes the largest proportion of emissions from the aviation industry (approximately 96 per cent). Emissions from airport operations in Norway are estimated at approximately 14,000 tonnes (less than 1 per cent in 2006),[6] while emissions from ground transport services to Norwegian airports are estimated at 67,000 tonnes (around 3 per cent) in 2007 (see Figure 15.2). However, emissions from airport operations and ground transport access are not recorded in the statistics of emissions from aviation, but are covered by other national statistics such as land transport.

Although it is evident from Figure 15.2 that some areas are more emission intensive than other, the three work groups in the project identified potential areas of improvement within aircraft engineering and maintenance, aircraft operations and ground transport access to the airports. Altogether, more than 50 improvements were proposed for the period up to 2020. In general the initiatives were based on known technology, anticipated technological developments and new initiatives emerged during the project. The most important initiatives have been developed and adopted. A prerequisite for the work groups was that all proposed initiatives had to be realistic and that the effect should be properly documented.

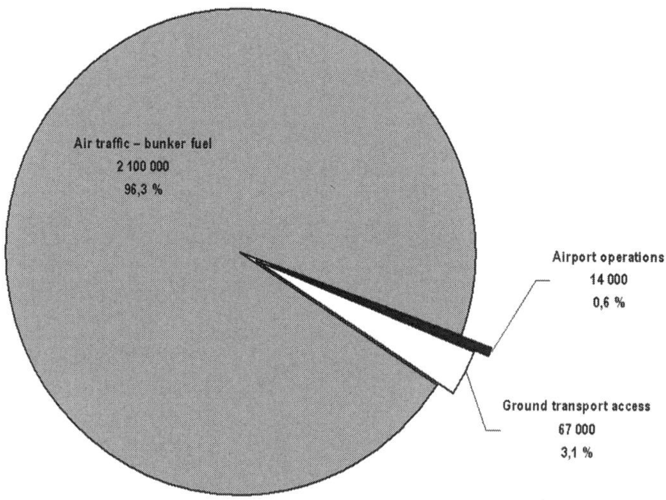

Figure 15.2 Greenhouse gas emissions from aviation in metric tonnes and as percentages, domestic and international (bunker fuel) (2005), airport operations in Norway (Avinor and OSL) (2006) and ground transport access in Norway (estimated) (2007)

6 Please note that most electricity in Norway is generated by hydro power, and that the number quoted is for the airport operator Avinor only.

Initiatives Related to Aircraft Engineering and Aircraft Operations

Initiatives related to aircraft engineering and operations will have the greatest impact on emission reductions up to 2020. Table 15.1 summarises the main groups of initiatives to be undertaken.

Initiatives on the existing aircraft fleet cover modifications such as the retrofitting of winglets, engine flushing, improvements to the airframe, optimising aircraft operations procedures and upgrading planning tools and computer systems. Due to fleet replacement, the effect of some of these initiatives will only apply in the short term, whereas others will also have an effect in 2020 and further on.

Aircraft fleet replacement is expected to produce a reduction in greenhouse gas emissions of around 25–30 per cent by 2020 as compared to 2007. Norwegian Air Shuttle is to replace all of its old Boeing 737 fleet with the latest generation technology (Boeing 737-800), whereas SAS Norway already had approximately 70 per cent of that aircraft type in its fleet during 2008. SAS is planning to replace parts of its aircraft fleet with the next generation of aircraft once these become available, probably around 2015. The CO_2 emissions from these aircraft are expected to be approximately 30 per cent lower per seat-kilometre than the current generation. In addition, SAS will be implementing other initiatives on its existing fleet, such as the installation of new seats, resulting in lower weight and greater seating capacity.

It is assumed that the overall effect of initiatives to reduce taxiing times and new fuel-efficient technology for aircraft movements at the airports will be between 3 and 7 per cent of total aircraft emissions. The high estimate takes alternative taxiing technology, e.g. wheel tugs, into account.

Table 15.1 Principal grouping of initiatives relating to aircraft engineering and operations, and emission reductions in Norwegian aviation up to 2020

Initiative	Emission reduction
Initiatives taken on existing fleet	5–1%
Fleet replacement	25–30%
Reduced emissions from taxiing	3–7%
Oslo ASAP and CDAs and CCDs at Norwegian airports	1–2%
Single European Sky (SES)	3–5%
Total	30–40%
Potential added impact from blend of fossil Jet A1 and biofuel	10–20%

The aviation authorities in 30 European countries are working on efficiency measures in European airspace (Single European Sky – SES). One objective of these is to reduce greenhouse gas emissions from an average flight by 10 per cent. It is anticipated that the emission reductions within Norwegian airspace will be between 3 and 5 per cent, mainly because of relatively little potential for optimising the use of this airspace compared to core areas in Europe.

In addition, the introduction of a new traffic management system in the eastern Norway region (Oslo ASAP) and optimised landings and departures at around 20 Norwegian airports are estimated to produce a reduction in emissions of 1-2 per cent.

Based on a total assessment of the impact of the aforementioned initiatives, the industry expects that a 30–40 per cent reduction in emissions by 2020 is realistic.

Finally, it is assumed that positive developments in the supply of alternative fuel and blending of carbon-neutral fuel could reduce emissions by between 10 and 20 per cent.[7] However, these estimates are uncertain and are only discussed here as an illustration of the potential for further reductions in greenhouse gas emissions from the industry, particularly after 2020. The industry's goal is to participate in trials of alternative fuels. Today, it is possible to have a blend of up to 50 per cent of synthetic fuel, which could also be produced from biomass.[8] The supply of alternative fuel, price and financial incentives will be crucial to the speed at which this can be introduced.

Initiatives Concerning Airport Operations

Greenhouse gas emissions from airport operations in Norway[9] are estimated to be approximately 14,000 tonnes per year. In general, it is anticipated that there will be a 25 per cent increase in greenhouse gas emissions in the period 2007–2020 if no efficiency initiatives are taken, because of traffic growth and expansions.

Avinor has calculated that it will be possible to substantially reduce greenhouse gas emissions from its own activities in 2020, even with the anticipated traffic growth. The main initiatives are outlined in Table 15.2.

The table indicates the potential for a 30–35 per cent reduction of emissions in all of Avinor's business operations. However, there is a degree of uncertainty linked to the impact of some of these initiatives. Avinor has therefore chosen to use as its basis an objective of emissions in 2020 being 25–35 per cent lower than they would have been if no initiatives had been taken at all.

---

7 Assumed reduction with 50 per cent blend and availability at a limited number of airports.

8 For example, Fischer Tropsch Bio-to-Liquid (BTL).

9 With regard to airport operations, only Avinor airports, including Oslo Airport Gardermoen, have been assessed in the project.

Table 15.2 Initiatives and potential reductions in emissions which Avinor control at the airports

Initiatives	Emission reduction
District heating and renewable sources of energy in own heat production	80–90%*
Reduction in own electricity consumption	25%
Reduction in fuel consumption in own vehicles	20–30%
Carbon-neutral fuel in own vehicles	80–90%*
Technological developments and initiatives related to business trips and journeys to/from work	25–30%
Carbon-neutral fuel in standby power	80–90%*
Potential for collective emission reductions	30–35%

* Assumes that adequate carbon-neutral fuel is available.

All large and medium-sized airports in Norway have hydronic heating systems, where most of the heat is currently generated in electric or oil boilers. Avinor's goal is to have heat production at these airports by 2020 based on carbon-neutral district heating, seawater heat pumps or biofuel (e.g. fired by woodchip and biogas), to ensure that between 80 and 90 per cent of total corporate heat production is based on carbon-neutral sources. In 2008, Avinor initiated a comprehensive energy conservation project with a target of 20 per cent reduction in electricity consumption during the course of a five-year period. In addition, a further reduction of 5 percentage points is anticipated during the period up to 2020.[10]

Replacement and modernisation of the stock of vehicles and other initiatives for reducing fuel consumption are expected to yield a fuel reduction of around 20–30 per cent during the period. In addition, Avinor is aiming for 80 per cent of the fuel consumed in the group to be carbon-neutral by 2020.

Oslo Airport has been offsetting the greenhouse gas emissions from its own operations from the financial year 2006 onwards by purchasing UN-approved emission rights.[11] Avinor has been offsetting emissions from its own operations on the other 45 airports in the network from the financial year 2008.

10 Emissions figures related to electricity consumption in 2006 are, in consultation with Det Norske Veritas (DNV), based on the assumption that less than 1 per cent of the electricity on the Norwegian market was imported. Among other factors greenhouse gas emissions related to electricity consumption are expected to vary from year to year depending on the precipitation situation in Norway.

11 The greenhouse gas audit has been verified by Det Norske Veritas (DNV).

Initiatives in Ground Transport Access

Ground transport access to airports causes considerable greenhouse gas emissions. The responsibility for good, eco-friendly ground transport to airports rests primarily with local and central government authorities and with the transportation companies concerned. However, Avinor supports the objective of public transportation increasing its share of the ground transport to the airports up to 2020, and will help facilitate this by available means.

The project estimated that in 2007, total emissions from ground transport to Avinor's airports were approximately 67,000 tonnes of CO_2 equivalents. The largest source of emissions is the private car. Among the major airports in Norway, the public transportation proportion is highest at Oslo Airport and lowest at Stavanger and Bergen airports.

Separate studies have been carried out concerning ground transport access to Oslo, Stavanger and Bergen airports. These indicate that it is possible to guarantee good public ground transport to the airports in the longer term and that the proportion of public transportation can be increased (Lian et al. 2007b; Asplan Viak 2007a, 2007b).

Table 15.3 shows estimated greenhouse gas emissions from Oslo, Bergen and Stavanger airports in 2007 and 2020. Ground transport access to these airports is responsible for approximately 70 per cent of total emissions from the total ground transport access to Norwegian airports. The table is based on public transportation proportions at these three airports of 62, 21 and 11 per cent respectively.[12] The table clearly illustrates the effect of increasing the proportions of public transportation to 70, 32 and 30 per cent, respectively, and of more eco-friendly car transportation for the three airports.

Table 15.3 Estimated CO_2 emissions (tonnes) from surface access to Oslo, Bergen and Stavanger airports in 2007 and 2020, with the corresponding objectives for public transportation proportions of 70, 32 and 30 per cent fulfilled

2007 CO_2 emissions				**2020 CO_2 emissions with fulfilled public transportation objectives**			
Oslo Airport	Bergen Airport	Stavanger Aiapirt	Total	Oslo Airport	Bergen Airport	Stavanger Airport	Total
32,700	8,200	5,500	46,400	31,300	7,600	4,700	43,600

12 The figures are based on Norway's National Survey of Travel Habits. Data are from the first half of 2007.

At Oslo Airport, a good, reliable range of train services with competitive prices is essential to achieving public transportation objectives. The most important initiatives in Bergen and Stavanger in the short term are an improved range of bus services (including new routes and increased frequency), but if the proportion of public transportation is to increase significantly, road improvements, bus lanes, new bus routes and possibly light rail systems will have to be implemented. To help facilitate this work, Avinor has established ground transport access forums at Stavanger and Bergen airports. In addition, Avinor will generally make arrangements for buses, including improved facilities, improved traffic information and marketing at the airport. Another example of work being done in this area is that the Airport Express train in Oslo has set up later departures for a trial period with financial support from Oslo Airport. Initiatives will also be taken to encourage the use of more climate-friendly fuel in the fleet of private motor vehicles. At present there are e.g. specific free parking spaces with charge-up areas for electric cars at Oslo Airport. Initiatives to be taken at airports to encourage a more eco-friendly stock of cars within the taxi industry will also be evaluated.

Traffic Forecasts and Greenhouse Gas Emissions in 2007–2020 If No Initiatives Are Taken[13]

Forecasts of traffic growth are important in any assessment of future greenhouse gas emissions. Traffic growth has varied historically and will vary considerably from one year to another during the period up to 2020. Up to 2008 the sector saw a period of strong growth, which was above the historical average and the expected long-term trend. It must, however, be emphasised that there is a great deal of uncertainty about any forecasts of traffic growth.

As regards the period up to 2020, forecasts indicate an average increase in annual traffic of 2.8 per cent at Norwegian airports (Lian et al. 2007a).[14] Annual growth for domestic flights is expected to be 1.9 per cent, international growth being 4.4 per cent. The leisure market is expected to show strong growth. This forecast formed the starting point for the calculations carried out in the project. However, with increases in direct scheduled international flights, estimates indicate that growth in passenger-kilometres from Norwegian airports could be around 75 per cent (Table 15.4) during the period 2007–2020.

13 Here, the term "no initiatives" should be interpreted as "no energy-efficiency initiatives", i.e. various technical, technological and operational initiatives that the aviation industry could implement. The project's task has been to assess the consequences and collate the initiatives that have been taken, based on existing technology, as well as to develop and promote new initiatives.

14 Calculated in terms of terminal passengers at Avinor airports, where domestic passengers count twice. If the number of journeys is used as a starting point, growth becomes just over 3 per cent per year.

Table 15.4 Traffic trends in passenger-kilometres and calculated emissions of CO_2 equivalents (not including helicopters) from bunker fuel sales with no energy-efficiency initiatives in 2007 and 2020

	1000 million passenger-km			Million metric tonnes CO_2 emissions no initiatives		
	2007	Growth to 2020		2007	Growth to 2020	
Domestic flights	5,8	1,6	28%	0,9	0,26	28%
International flights	10,2	10,4	102%	1,12	1,13	101%
Total	16,0	12,0	75%	2,02	1,39	69%

Source: Based on Lian et al. (2007a).

Table 15.4 also shows estimated emissions in 2020 if no industry-specific initiatives are taken. This includes an extrapolation of current emissions per passenger-kilometre and anticipated traffic growth. The growth in passenger numbers and emissions is least for domestic and greatest for international traffic. The main reason for the calculated strong growth in emissions for international flights is the increase in direct scheduled flights to European and intercontinental destinations. This means that the strongest increase will be in the proportion of emissions from international traffic, which is included in the bunker fuel statistics, and therefore within Norwegian aviation's "responsibility", as laid out in the project.

The Effect of Initiatives Taken by the Industry in 2020

Figure 15.3 summarises the effect of the initiatives outlined and discussed in the project. The traffic forecasts used as the basis for the calculations estimate a growth in emissions of 69 per cent of bunker fuel sales with no initiatives being taken. The initiatives outlined however are expected to result in a reduction in emissions of 30-40 per cent, which means that greenhouse gas emissions from the industry in 2020 should be 0–15 per cent higher than they were in 2007. In other words, emissions in 2020 will be at a level of between 2.0 and 2.3 million tonnes, as compared to 2.0 million tonnes in 2007.

Because considerably lower growth is expected in the domestic market (28 per cent), the calculations indicate greenhouse gas emissions in 2020 that are between 10 per cent and 22 per cent lower than in 2007, or emissions of between 0.73 million tonnes and 0.81 million tonnes in 2020, as compared to around 0.9 million tonnes in 2007.[15]

15 The numbers do not cover helicopter traffic.

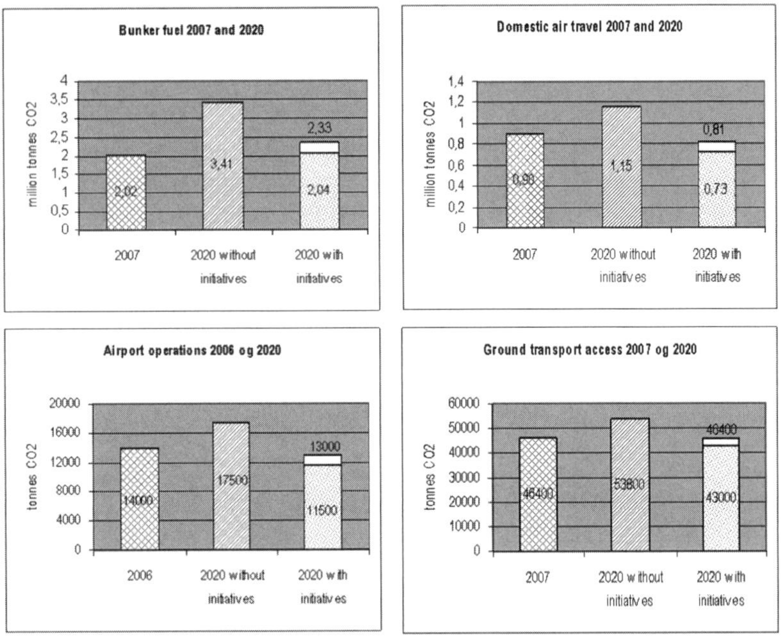

Figure 15.3 A summary representation of greenhouse gas emissions in 2007 (2006 for airport operations) and 2020, with and without energy efficiency initiatives

Note: Different scales on the y-axis.

Regarding airport operations, the calculations indicate that greenhouse gas emissions will be between 8 and 18 per cent lower in 2020 than in 2007, even with considerable growth in traffic. This equates to emissions of between 11,500 and 13,000 tonnes in 2020.

Avinor will be helping to reduce the environmental load from ground transport by means of the instruments which the group has at its disposal. Lian (2007b) and Asplan Viak (2007a, 2007b) indicate that, in total, emissions from ground transport access to Oslo, Bergen and Stavanger airports may be 19 per cent lower in 2020 than they would have been if no initiatives were taken. If the public transport objectives are achieved, emissions are estimated at between 43,000 and 45,700 tonnes in 2020, as compared to 46,400 tonnes in 2007.

The initiatives outlined in the project will be implemented up to 2020. The airlines and Avinor are responsible for implementing the initiatives in accordance with their areas of responsibility. In three-year intervals the industry will evaluate the impact of the initiatives. There will necessarily be a degree of uncertainty connected with forecasts up to 2020. The most important factors that might weaken/strengthen the impact of the industry's initiatives are traffic growth, technological developments, the availability of alternative fuels and the general competitive situation.

Conclusion

The project on sustainable and socially beneficial aviation in Norway showed that it is possible to combine aviation growth and emission abatement. Combining statistics of traffic growth with emissions, and extrapolating the two variables in accordance with the traffic forecasts and the projected effect of the emission reduction initiatives gave an interesting picture.

Figure 15.4 shows the relation between traffic growth[16] and CO_2 emissions from bunker fuel sales during the period 1990 to 2020. During the period 1990 to 2007, there was an average annual growth in air traffic from Norwegian airports and in CO_2 emissions of 4.6 and 3.6 per cent respectively. The impact of the initiatives the industry has taken will strengthen this trend. Emissions in 2020 will thus be at around the same level as in 2007 or a little higher, despite continued significant traffic growth.

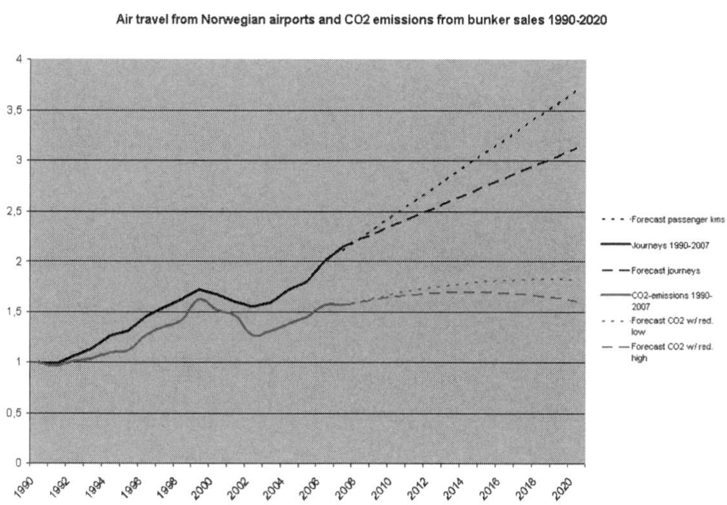

Figure 15.4 Air travel and CO_2 emissions from bunker fuel sales during 1990–2007, and forecasts of traffic growth and greenhouse gas emissions up to 2020 (indexed)

16 The figure only shows the historical trend in the number of journeys because no historical data exist for passenger-kilometres. The forecast gives a stronger growth in passenger-kilometres than for journeys.

Figure 15.5 shows the relation between domestic traffic growth and CO_2 emissions in Norway during the period 1990 to 2020. It indicates that the industry's environmental performance is improved in the domestic market during the period up to 2020 as compared with the historical trend. Because the growth in domestic air traffic is expected to be lower than for international traffic, the initiatives taken by the industry will lead to emissions being 10–22 per cent lower in 2020 than in 2007.

The expected consequences of climate change represent important challenges. The aviation industry therefore has to bear responsibility for reducing its impact on the climate. This must be accomplished by the industry covering the costs of the environmental disadvantages that are being inflicted upon society (the "polluter pays" principle) and by implementing initiatives to reduce emissions. With the project initiated in Norway, the industry outlined tangible initiatives that will make considerable improvements to the industry's environmental performance. The most important initiatives have already been developed and adopted, and decisions will be made on implementing the rest once the technology becomes available.

Due to its geography and topography, Norway is dependent on air travel. Thus, the Norwegian aviation industry concluded that a scenario with an increase in emissions proportional to the growth in air traffic is unacceptable because of the environmental consequences. On the other hand, a controlled restriction of air traffic growth is also unacceptable because of the negative societal consequences.

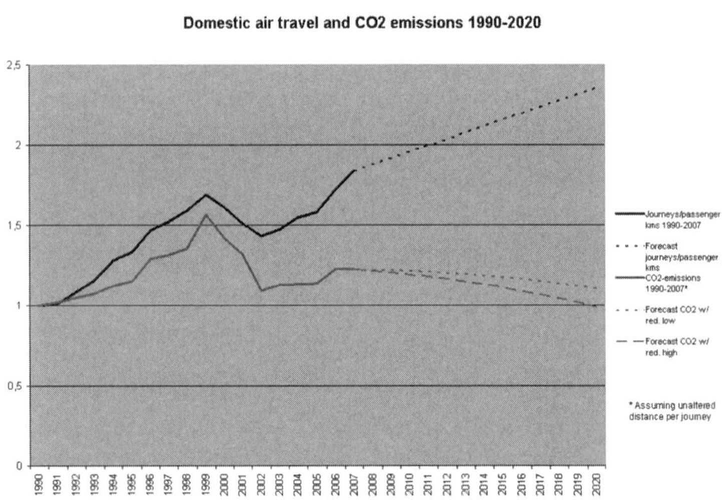

Figure 15.5 Domestic air travel and CO_2 emissions from bunker fuel sales during 1990–2007, and forecasts of domestic traffic growth and greenhouse gas emissions up to 2020 (indexed)

A "third way", where emissions are stabilised and decoupled from traffic growth, is thus described above. However, bearing in mind that the IPCC maintains that emissions in the OECD countries must be reduced by maybe as much as 80 per cent by 2050, the initiatives discussed in this chapter are only part of the first leg in a 40-year intense relay race.

References

Asplan Viak (2007a), Tilbringertjenesten til Stavanger Lufthavn, Sola [Ground Transport Access to Stavanger Sola Airport]. Issue: 01, 30 November 2007.

Asplan Viak (2007b), Tilbringertjenesten til Bergen lufthavn, Flesland [Ground Transport Access to Bergen Flesland Airport]. Issue 02, 13 December 2007.

Avinor (2008a), *Bærekraftig og samfunnsnyttig luftfart. Prosjektrapport* [Sustainable and Socially Beneficial Aviation. Project Report] Oslo, Avinor, 30 January 2008 (available at http://www.avinor.no or from the authors).

Avinor (2008b), *Aviation in Norway. Sustainability and Social Benefit. Project Report Summary*. Oslo, Avinor, 30 January 2008 (available at http://www.avinor.no or from the authors).

Forster, P.M.D., K.P. Shine and N. Stuber (2007), "It is premature to include non-CO_2 effects of aviation in emission trading schemes", *Atmospheric Environment*, 41, 3941 (corrigendum to 40, 1117–21, 2006).

Fuglestvedt, J., T. Berntsen, G. Myhre, K. Rypdal and R. Bieltvedt Skeie (2008), "Climate forcing from the transport sectors", *Proceedings of the National Academy of Sciences*, 105, 454–8.

IPCC (1999), *Aviation and the Global Atmosphere*. A special report of IPCC working groups I and III. J.E. Penner, D.H. Lister, D.J. Griggs, D.J. Dokken and M. McFarland (eds).

Lian et al. (2007a), "Bærekraftig og samfunnsnyttig luftfart" [Aviation – Sustainability and Social Benefit]. *TØI Report 921/2007*. Oslo, TØI.

Lian et al. (2007b), "Transport til/fra Oslo Lufthavn i et langsiktig perspektiv" [Transport to/from Oslo Airport in a long-term perspective]. *TØI Report* 902/2007. Oslo, TØI.

Stordal, F. et al. (2005), "Is there a trend in cirrus cloud cover due to aircraft traffic?", *Atmospheric Chemistry and Physics*, 5, 2155–62.

Chapter 16
A Replacement Sub-20 Seat Aircraft for Remoter Regions Operations

Andy Foster

This chapter will look at the challenge of designing and producing aircraft suitable for transporting passengers and goods on some of the most difficult routes in the world. The key aircraft which have been used, and are still being flown in such areas of the world, will be discussed along with those designs which are planned for entry into service in the near future. Finally a new aircraft design is proposed based on research work carried out by a postgraduate student at Cranfield University (Quéméneur, 2008).

The sub-20 seat category of aircraft will be considered since not only do such aircraft operate on many remoter regions' services, but there are issues of certification (this will be covered later) and tax issues which make such aircraft attractive to operators. In terms of tax benefits such aircraft are exempt from Air Passenger Duty in the UK and also will be exempt from inclusion in the soon to be implemented Emissions Trading Scheme in the European Union.

At present there are over 3,000 sub-20 seat aircraft in service worldwide operating in a variety of roles and a wide range of climate and infrastructure conditions. The most popular turboprop powered types that are currently flown are discussed below.

Current Aircraft in Service

De Havilland DHC-6 Twin Otter

To many this is the ubiquitous type of aircraft in this category. Initially developed from the single-engined DHC-3 Otter aircraft, the first DHC-6 flew in 1965. Funded by the Canadian government, the aircraft was specifically designed for Short Take-Off and Landing (STOL) operations. To assist with this the aircraft was equipped with double slotted flaps and ailerons which work in unison with the flaps to improve the landing performance. The aircraft was powered by two Pratt & Whitney Canada PT6A turboprops – uprated versions of which have been used on all subsequent aircraft.

The Twin Otter soon found a ready market in the civil and military field in a variety of roles, from scheduled passenger service to paradropping and special operations. The intial variant, the Series 100, entered service in 1966. The first major change for the aircraft, the Series 200, incorporated an extended nose, including a bigger baggage area; and with a larger storage compartment at the rear of the aircraft it meant that overall the luggage space on board was greatly increased. The first Series 200 aircraft entered service in April 1968. In total 231 Series 100 and 200 aircraft had been built before the final production variant built by de Havilland Canada, the Series 300, entered service in 1969. This variant featured more powerful engines and heavier design weights, which gave it the ability to carry up to 20 passengers.

Over the years the aircraft has been equipped with wheels, but also float plane and ski versions have been produced and it is testament to the design and production standards of de Havilland Canada that the aircraft is able to operate in a wide variety of climates, from the frozen wastes of Antarctica to the tropical heat of Papua New Guinea.

An example of the importance of the aircraft to remoter regions in Europe is shown by the use of the aircraft by Widerøe's Flyveselskap in Norway. Widerøes had been operating services since the 1930s when in the late 1960s the Norwegian government and Widerøe proposed to build a network of 19 800m STOL airfields in northern and western Norway designed around the Twin Otter. Widerøe took delivery of their first Twin Otter in 1967 and in the 1970s was one of the largest operators of the type. Their last Twin Otter left service in 2000, by which time they had been replaced by the larger and more comfortable Dash 8 100. The importance of the type to Norway is indicated by the presence of the first Twin Otter operated by Widerøe in the Norwegian Aviation Museum in Bodø.

By 1988, de Havilland Canada had been taken over, along with Canadair, by Bombardier and merged into Bombardier Aerospace and a decision was made to halt production of the Twin Otter, by which time some 844 of the type had been produced, of which 115 were Series 100, 115 Series 200 and 614 Series 300. The support for the aircraft remained with Bombardier until 2005 when a Canadian aviation support company called Viking Air took over.

In early 2009 there were over 550 Twin Otters listed as being active, with an average age of 33 years. The oldest aircraft in civil service is the ninth production aircraft, in service with Sun Air (Pacific) based in Nadi, Fiji. This aircraft is now over 42 years old. Even the youngest Twin Otter in service, serial number 842, is 20 years old and is operated by ConocoPhillips in Alaska.

While older aircraft generally have drawbacks in terms of higher maintenance costs and poorer reliability, the lack of a suitable replacement aircraft has meant that the value of the Twin Otter has remained high in the market place. Table 16.1 shows the geographical distribution of Twin Otters in commercial service in 2008 and Figure 16.1 depicts one of the amphibian aircraft operated by AirSeaLines of Greece.

Table 16.1 Geographical distribution of Twin Otters in commercial service in 2008

Africa	Asia, Australasia and Middle East	Europe	North/South America
50	106	29	142

Source: *Flight International* (2008).

Figure 16.1 Twin Otter Series 300

Source: Photo courtesy of Eryl Crump.

Dornier 228

Developed in the early 1980s, the Dornier 228 resulted from German government funded research using the Dornier 28 Skyservant and its successor, the Do 128. The new aircraft featured an unpressurised cabin and a novel wing design known as TNT (*Tragflügel neuer Technologie*). Powered by two Garrett TPE 331 turboprop engines, the type first flew in 1981 as the Series 100 seating 15 passengers, shortly followed by the Series 200 seating 19 passengers. In 1989 licence production of the type started at the Hindustan Aeronautics Ltd (HAL) factory in Kanpur, India.

The aircraft has seen service in a number of roles around the world, not only operating scheduled passenger services, but also fisheries patrols. An example of the use of the Dornier 228 in remoter regions operations is shown by the operation by SATA in the Azores. Their aircraft was delivered in 1993 and has been used on services to the island of Corvo, which has a runway of 800 m (Figure 16.2).

In 1996 Fairchild of the US bought Dornier to form a new company – Fairchild Dornier – which looked to develop new and larger turboprop and turbofan powered aircraft, including the 428, 528 and 728 types. With this in mind the company ceased production of the 228 in 1999, by which time some 244 aircraft had been produced in Germany. Unfortunately in 2002 Fairchild Dornier entered administration and eventually was split up, with RUAG, a Swiss-based company, taking on the air services arm of the company and since then it has continued to support the 228 aircraft in service.

In early 2009 there were around 230 of the type in service, the eldest being the 26-year-old third production series 100 operating with Businesswings of Kassel-Calden, Germany. The average age of all Do228s still in service is 19 years. Table 16.2 shows the geographical distribution of Dornier 228s in commercial service in 2008.

Figure 16.2 Dornier 228

Source: Photo courtesy of Eryl Crump.

Table 16.2 Geographical distribution of Do228s in commercial service in 2008

Africa	Asia, Australasia and Middle East	Europe	North/South America
21	48	18	32

Source: *Flight International* (2008).

Harbin Y-12

The Chinese Y-12 aircraft was developed from its predecessor the Y-11, which was a high-wing twin piston engined aircraft which had first flown in 1975. The Y-12 was launched in 1982 and again was a high-wing design – though featuring a more advanced, supercritical wing and powered by two Pratt & Whitney PT6A-11 turboprops; this first version, the Y-12-I, had seating for 17 passengers. It was followed by the Y-12-II in 1984, which used more powerful PT6A-27 engines and featured the removal of leading edge slats and an increased Maximum Take-off Weight (MTOW). The next variant produced was the Y-12-IV, which entered service in 1988 and featured seating for 18–19 passengers. In the late 1990s Canadian Aerospace Group set up a joint venture with the Chinese to complete Y-12s in Canada, and that variant was to be known as the Twin Panda. It was based on the Y12-IV and featured a more powerful engine (the PT6A-34), upgraded avionics, a new interior and strengthened landing gear (Figure 16.3). A number of orders were received; however, in the end the project was abandoned.

It is believed that over 130 Y-12s have been produced and have served mainly in China itself and other Asian states. Unfortunately there is no data available to indicate their exact whereabouts.

Figure 16.3 Harbin Y-12-IV
Source: Photo courtesy of AVIC.

Let 410

The Let 410 was initially launched by the Czech aircraft manufacturer Let Kunovice in the 1960s, partly as a resonse to a requirement from Aeroflot for a replacement for the Antonov 2 biplane. Christened the Turbolet, the high-wing 15–20-seater aircraft was due to be equipped by locally produced Walter M601 turboprop engines. However development issues meant that the first Let 410, which flew in 1969, was powered by Pratt & Whitney Canada PT6A-27 engines and the first aircraft to enter service the following year were also powered by those engines. Later in 1970 the development issues with the Walter engine were overcome and all subsequent production aircraft have been powered by the locally produced engine. Over the years the aircraft was developed with improved engines and other aerodynamic improvements to optimise its STOL performance. As well as being produced as a regional airliner, the Let 410 has been adapted to operate as an air ambulance and paratroop transport aircraft (Figure 16.4). The current production variant is the Let 410 UVP-E20.

It is estimated that over 1,000 Let 410s of all variants have been produced during the last 40 years. The oldest still in service is operated by Bluebird Aviation of Kenya, and is 39 years old. The average age of the fleet is 23 years. Table 16.3 shows the geographical distribution of Let 410s in commercial service in 2008.

Figure 16.4 Let 410

Source: Photo courtesy of Eryl Crump.

Table 16.3 Geographical distribution of Let 410s in commercial service in 2008

Africa	Asia, Australasia and Middle East	Europe	North/South America
57	20	123	72

Source: *Flight International* (2008).

Other Aircraft

There are a number of other aircraft of similar size and performance to the Twin Otter, Do228 and Y-12, including the GAF Nomad (Figure 16.5), Shorts Skyvan (Figure 16.6) and PZL M-28 Skytruck (Figure 16.7). Table 16.4 provides some technical data in respect of each of these aircraft types.

Figure 16.5 GAF Nomad

Source: Photo courtesy of Edwin Yii.

Figure 16.6 Shorts Skyvan

Source: Photo courtesy of Eryl Crump.

Figure 16.7 PZL M-28 Skytruck

Source: Photo courtesy of Jaap Niemeijer.

Table 16.4 Comparison data of existing aircraft

Aircraft	MTOW	OEW	Engines	Number in service (estimate for all variants)
Twin Otter 300	5670 kg	3363 kg	2 × PT6A-27	550
Do228-200	6350 kg	3260 kg	2 × TPE331	230
Harbin Y-12-IV	5670 kg	4000 kg	2 × PT6A-27	130
Let 410	6600 kg	4225 kg	2 × Motorlet M601E	780
PZL M-28	7000 kg	3900 kg	2 × PT6-65B	120
Shorts Skyvan	5670 kg	3330 kg	2 × TPE331	35
GAF Nomad	4268 kg	2377 kg	2 × Allison 250BC	86

MTOW = Maximum Take-off Weight; OEW = Operating Empty Weight.

Sources: Airliners.net (2009); Quéméneur (2008).

Aircraft Due to Enter Service in the Near Future

Viking Twin Otter 400

As mentioned earlier, in 2005 Viking Air took over the support function for the Twin Otter and in so doing also took on the design authority. Given the lack of a ready replacement aircraft for the Twin Otter they decided to revisit the design to see if they could develop a new variant. By 2006 they were convinced that there was sufficient market demand to go ahead with the project, and at the Farnborough Air Show they formally announced their intention to proceed with the Twin Otter 400 as they called it (Figure 16.8). Overall the design is the same as the Series 300, but Viking have made a number of changes. While the primary structure of the aircraft is unchanged, some secondary structures (for example, the radome and door) are being made from lighter composite materials. The engines to be used are new and more powerful PT6-34 turboprop engines which will give the aircraft better airfield and fuel performance. The Series 400 has a modern glass cockpit and updated avionics systems compared to the Series 300.

The first Twin Otter 400 flew in October 2008. So far Viking have orders for some 40 of the new type, including from long-standing Twin Otter customers Zimex Aviation of Switzerland and Maldivian Air Charter of the Maldives. They plan to produce 200 over the next 10 years based on a market forecast of 400 for this size aircraft.

Figure 16.8 Viking Twin Otter 400

Source: Photo courtesy of Viking Air Ltd.

RUAG 228 New Generation (NG)

When RUAG acquired the aircraft services division of Fairchild Dornier in 2002 it also took on the rights and licences as original manufacturer of the type. Following the completion of studies into the market the board of RUAG were convinced that there was a case for developing a new variant of the 228, and in 2007 it announced that it was going to develop the RUAG 228NG (New Generation) – Figure 16.9. As with the Twin Otter 400, the RUAG 228NG will be equipped with the latest digital avionics and a glass cockpit. The engines chosen are Garrett TPE331-10s. The propeller is a new five-bladed design which will help to reduce noise, as well as improving the airfield performance capabilities of the aircraft. RUAG, through a partnership agreement with HAL, is reactivating the production facility at Kanpur where the fuselage, wings and tail will be produced. Final assembly and internal fitting of the aircraft will be completed at the former Dornier facility at Oberpfaffenhofen, near Munich in Germany.

The first NG aircraft flew in September 2008. At the time of writing RUAG has announced orders for five of the type from Mexico, plus additional aircraft from undisclosed customers in Australia and Japan. Total orders for 15 aircraft have been received by mid-2009.

GECI Skylander

GECI is a French-based industrial company which in 2001 produced a design for a new aircraft. In overall design the Skyland SK-105 is a high-wing, twin-engined aircraft. Initially the plan was to produce the aircraft in Portugal, but in September 2008 the company announced that it would develop the aircraft at a new facility to be developed at a former NATO airbase at Chambley-Buissières (30 km from Metz). Current plans are for certification and entry into service of the type in 2011.

Figure 16.9 RUAG 228NG

Source: Photo courtesy of RUAG Aerospace Services GmbH.

The type is being offered in a number of variants, including: 19-seat passenger layout; cargo layout (with ability to accommodate three LD3 containers or four 1.4 m × 1.8 m pallets); combi (mix of passenger seats and space for cargo); air ambulance and paratroop transport (Figure 16.10). By mid-2009 the company had claimed orders for 15 aircraft from ACT Airlines Group of Turkey, for freight operations, and four aircraft from Transasia in Malaysia. The company has plans for building over 1,500 of the type between 2012 and 2028.

In terms of dimensions and weights the Skylander is longer and heavier than many of its competitors – mainly due to the requirement to carry 3.3 tonnes of payload – with a design MTOW of around 8,400 kg (nearly 3,000 kg more than the Twin Otter 400). In turn this means that the Skylander requires more powerful engines and to this end is to be equipped with two PT6-65Bs producing 1,100 shp each.

Figure 16.10 Skylander SK-105

Source: Artist's impression courtesy of GECI International SkyAircraft.

Harbin Y-12F

The last aircraft to be considered in the new aircraft category is the latest variant of the Y-12, known as the Y-12F. In fact it is almost a completely new design, with a wider fuselage, rectractable landing gear and more powerful Pratt & Whitney PT6A-34 engines and five-bladed Hartznell propellers. Designed for passenger or cargo use the aircraft features advanced avionics and cockpit. While much of the structure will be conventional aluminium alloy, the aircraft will incorporate composites making up some 7–10 per cent of the overall structure. Launched in 2006 and displayed as a model at the Paris Air Show in 2007, the Y-12F is due to fly in 2010 and enter service in 2011 (Figure 16.11).

Figure 16.11 AVIC Y-12F
Source: Photo courtesy of AVIC.

Proposed New Aircraft Design

This section will describe the design process for a completely new aircraft designed specifically to meet the needs of operators in remote regions. The work was carried out by Alexandre Quéméneur for his Individual Research Project towards his MSc in Aerospace Vehicle Design in 2007/08 (Quéméneur, 2008). The aircraft design was named the ALX-190 Lion Cub by the student.

Before designing a new aircraft it is important to set out the key requirements for the aircraft in terms of its performance (for example the length of runway required to take off and land), the amount of payload that can be carried and the range that is achievable taking into account the operating conditions it will face. It is also crucial to consider the cost of designing and manufacturing the aircraft and also the cost of operating it. Also, designers have to consider the airworthiness requirements for the aircraft. For "Commuter Class" aircraft with up to 19 seats the key airworthiness standards are those laid down by the US Federal Aviation Administration in Federal Airworthiness Regulation FAR 23 and by the European Aviation Safety Agency in Certification Standard CS 23. Some other countries do have their own airworthiness standards, but generally they follow the FAR or CS regulations. These airworthiness requirements are based on years of civil aviation experience and aim to ensure that new aircraft are safe to operate.

Once the requirements for the aircraft have been defined the actual design process is an interative one. The stages described by Howe (2000) as *Feasibility Study* through to the *Project Process* are also defined as the *conceptual design phase* (Roskam, 1990). For an aircraft manufacturer the conceptual design phase requires relatively small investment since most of it is designers' time and effort, and more and more these days is in computer power to generate data and results rather than the need for testing or manufacturing equipment. However at the same time it must be noted that decisions taken at these stages will have a major influence on the programme as a whole and hence they should not be taken lightly. Aviation history has shown that getting this phase wrong can lead to aircraft which are not commercially successful.

Following the conceptual design phase comes the preliminary design, where more detailed work is carried out on the key components of the aircraft and the systems on board. Roskam considered that the conceptual and preliminary design phases taken together account for around 85 per cent of the life cycle cost of a new aircraft. If a decision is taken to continue with the design at this point then the subsequent phases will be detail design and development followed by manufacturing and, hopefully, acquisition by customers. During the lifetime of an aircraft modifications will be made in terms of new variants, as we have seen already with the current aircraft in service, and also changes to meet airworthiness or operational requirements. In the meantime the aircraft manufacturers' designers will be continually looking at new designs for aircraft to replace those in service.

Requirements for the ALX-190

The following sets out the specific requirements defined for a new aircraft for serving remoter regions, based on a survey of the characteristics of the current aircraft in service defined earlier in this chapter.

Layout and Payload

- 19 passengers, aircraft certificated under FAR 23/CS 23 requirements
- Maximum payload (cargo configuration): 2,500 kg
- 3 industry standard LD3 containers to be carried in cargo configuration
- Aft cargo door
- Large range of centre of gravity position (for carriage of loose cargo)
- Alternative roles (cargo/fire fighting/seaplane/medical evacuation/troops and military transport/paratroops/maritime patrol, search and rescue/bush hospital/private). All alternative roles to be taken into account in the design of the baseline to reduce development and production cost.

Field and Flight Performance

- Take-off and landing rolling distance: 350 m (at MTOW, under ISA conditions at an airfield elevation of 2,500 ft)
- Maximum ceiling: 12,500 ft (non-pressurised cabin)
- Maximum cruise speed: 200 kts at an altitude of 9,000 ft
- Design mission: 19 passengers (payload 4173 lb/1875 kg) over 400 nm at a speed of 180 kts at an altitude of 9,000 ft.

In terms of structures and systems drivers, simplicity, ease of maintenance and robustness are the main design drivers.

Structures

- Primary structures: conventional aluminium alloy
- Secondary structures and flaps/ailerons/rudder: possibly composite to reduce weight
- Structural load (manoeuvre limit load): +3.8/-1.5
- Minimum number of structural components to reduce maintenance cost.

Systems

- Systems and flight controls: mechanical
- Cost drivers
- Low acquisition cost: maximum $4 m (based on 2007 prices)
- Low cost of operation: operating costs per seat mile less than both Twin Otter 400 and Skylander 100
- Production site(s) of the aircraft: dollar zone/emerging countries.

Other Drivers

The aircraft needs the ability to operate in extreme climate conditions (in terms of humidity and temperature), from unprepared runways and airfields located close to difficult terrain, in remote areas where spare parts availability may be a problem.

Airworthiness Requirements for the ALX-190

The aircraft was designed under FAR 23/CS 23, since the type has "a seating configuration, excluding pilot seats, of 19 or less and a maximum certificated weight of 8,600 kg or less" (Stinton, 2001). For military variants, the ALX-190 would have to comply with the particular military airworthiness standards required by the customer.

Having defined the requirements and the airworthiness standards to follow, the conceptual design process begins by defining the overall configuration of the aircraft.

Engine and Powerplant Installation

Since the ALX-190 is designed under FAR 23/CS 23 it requires at least two engines. Most aircraft in this size category, as we have seen earlier, have their engines mounted on the wings. This solution benefits both structurally, in that it leads to a reduction in wing bending moment, and also operationally since it means that the engines are accessible for maintenance. Other engine configurations are possible, including tail mounting (as with the B-N Trislander) or push-pull configuration (as with the Cessna 337 Skymaster), but these more unusual solutions do lead to challenges such as centre of gravity position and engine accessibility (Figure 16.12).

Having defined the location of the powerplant it is then necessary to select the particular type of engine to be installed. For the ALX-190 the student considered three options: piston, turboprop or turbofan engines. For cost and operational reasons (taking into account the aircraft requirements and required cruise speed), the turbofan was not considered a viable solution. Hence the choice fell to either piston or turboprop engines. Considering the power requirement for the aircraft and the operational conditions it would face (in terms of humidity and temperature), the turboprop engine was considered the best solution. This choice is also backed by the fact that in some parts of the world, for example in Africa, there is a general lack of Avgas 100LL gasoline (used for piston engines), while Jet A (used for turboprops) is largely available.

Based on the performance requirements the power required for each of the ALX-190 engines was calculated to be 746 shp. The most widely used engine in this class, as seen earlier, is the Pratt & Whitney Canada PT6, which has a good reputation for reliability and low fuel consumption, and with customers based around the world spare parts are relatively easy to obtain – a critical issue for an aircraft serving remote communities. The specific variant of the engine selected for the ALX-190 was the PT6-34 with a rating of 750 shp; this variant of the engine powers the Viking Twin Otter 400. The propeller selected for the ALX-190 was a four-bladed design.

Figure 16.12 BN Trislander, tail-mounted engine and Cessna 337 push-pull configuration

Sources: BN Trislander, photo courtesy of Eryl Crump and Cessna 337 courtesy of Matt Willems.

Fuselage Layout

The fuselage provides the volume to carry payload (passengers and/or cargo) while at the same time assuring the overall structural integrity of the aircraft. Since the ALX-190 was designed with flexibility in mind, the fuselage design needed to accommodate different types of loads. As with the aircraft described earlier, a non-pressurised cabin was chosen for the ALX-190. This has advantages in that it means that doors and windows can be designed to be larger than is possible for a pressurised cabin, although the drawback is that the aircraft in some cases will have to fly through the weather, leading to a less pleasant passenger experience.

Taking into account the cargo-carrying requirements for the aircraft, it was decided to incorporate an aft, ramp-style cargo door and to adopt a rectangular shape for the fuselage. This option is cheaper in terms of manufacturing. As previously mentioned, a maximum of 19 seats was decided on. The seating configuration selected for this was the classic three-abreast with an offset aisle with a seat pitch selected of 31 inches, which is more generous than offered on many low-cost or even network carriers. To comply with FAR 23/CS 23 requirements, three emergency exits were required and these locations are shown in the fuselage layout in Figure 16.13. With a cabin length of 5.7 m the ALX-190 was designed to carry three LD3 containers.

Wing Configuration

The position selected for a wing has an impact on the layout of the fuselage. Aircraft in this class, for performance and operational reasons, generally have a "high-wing" configuration where the wing is above the fuselage. This helps the wing generate more lift, but also keeps the engines away from the ground – particularly important when considering operations at airfields with rough surfaces or, in the case of a seaplane/amphibian variant, keeping the engines clear of water.

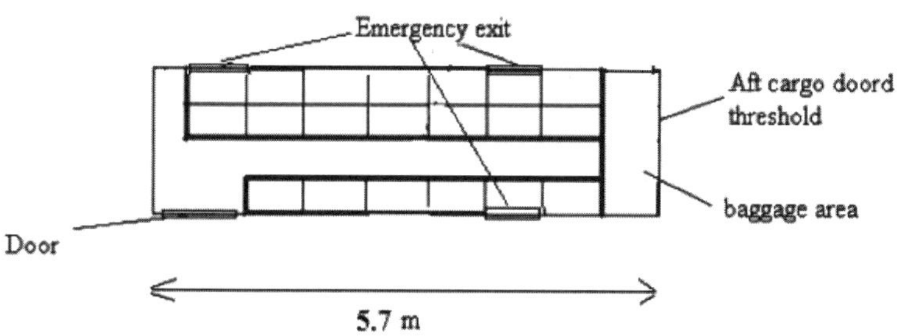

Figure 16.13 Cabin layout of the ALX-190

Source: Quéméneur (2008).

This configuration was also chosen for the ALX-190. In terms of structure for supporting the wing then a "braced wing" or "semi cantilever wing" are the two choices. The braced wing (as shown on the Twin Otter) means that the wing itself does not have to be as strong, and hence heavy. There is a drag penalty for such a configuration but for the ALX-190 this was felt to be more than compensated for by the weight benefit, and hence a braced wing was selected.

In terms of the wing planform, or shape, then a purely rectangular (or constant chord) wing, where the ribs are all the same, is preferable from a manufacturing perspective. However such a planform wing generates more drag compared with an elliptical planform wing, estimated at around 7 per cent more drag (Raymer, 2006). For the ALX-190 a compromise was chosen where the wing would be tapered, and hence the wing tip chord would be approximately half the chord at the wing root. Compared to the ideal elliptical wing (which would be costly to build) this would incur only a 1 per cent drag penalty.

To help the aircraft achieve its strict take-off and landing requirements flaps were incorporated; but to avoid added weight, complexity and hence cost, leading edge devices/slats were not felt to be required.

Tail and Fin

With the exception of a small number of aircraft designs (such as the Shorts Skyvan and Skytruck, which featured twin fins), the traditional configuration of the tail of aircraft in this class consists of a horizontal stabiliser and a single fin. Since the ALX-190 was designed with an aft ramp-style cargo door and has a high-wing configuration, a high-mounted tailplane was initially proposed. However, this configuration leads to at least two main drawbacks: increased structural weight and unfavourable behaviour in stalling conditions. Hence, it was finally decided to install the tailplane at the root of the fin. This configuration can lead to problems in spin recovery and to alleviate this problem a small dorsal fin was added in front of the fin itself.

Landing Gear

The landing gear of an aircraft designed to operate in remoter areas with unprepared runways is a critical part of the design. While having landing gear which can retract is useful from a drag perspective, it does add complexity in terms of manufacturing components and maintenance requirements and weight to an aircraft. Hence, as with a number of similar aircraft it was decided that the ALX-190 would have a fixed landing gear, i.e. a single nose wheel and main landing gear comprising single wheels under the port and starboard wings. The tyres required a lot of attention since they are critical to the ability to operate from unpaved runways.

Systems and Avionics

Key systems requirements for the aircraft were simplicity and robustness.

Environmental control system (ECS) Since the aircraft is unpressurised the ECS would be purely an air ventilation system (with outside air mixed with bleed air from the engine, fed through a system of filters and vanes).

Ice protection system (IPS) With turboprop engines it is essential that ice is prevented from entering the engine. Given that the ALX-190 was designed to operate in extremely cold conditions, e.g. northern Canada, an effective ice protection system was incorporated to prevent the build up of ice on leading edges, engine intakes and on the propellers. While pneumatic systems are common, it was decided to install an electrical IPS to reduce weight.

Fuel system The fuel system was designed to be simple and compliant with FAR 23/CS 23. The range performance of the aircraft, combined with its fuel consumption, led to the requirement for an overall volume of 54 cu ft/1.54 m^3 to provide capacity for 410 US gallons/1,200 kg.

Secondary power To reduce weight and complexity it was decided that hydraulic and pneumatic power were not essential and that the only secondary source of power would be electrical, produced by generators on the engines for the aircraft's instruments and for the IPS mentioned before.

Flight control system A traditional, but practical, approach was taken with regard to the flight control system in that a manual and totally mechanical system of rods, pulleys and cables was selected.

Cockpit and avionics While other parts of the aircraft will have been seen to be almost retrospective, it was decided when it came to the cockpit and avionics fit that a glass cockpit was justified by its high reliability and reduced weight, as well as its ease of use for pilots and engineers.

Weather radar will not be included in the standard fit, but will be available as an option. Moreover, special variants (military/surveillance/patrol) could integrate more avionics instead of payload.

Preliminary Design Phase

Having completed the conceptual design phase, the next stage in designing an aircraft is the preliminary design phase where more detailed work is carried out to ensure that the aircraft components meet the strict performance and airworthiness requirements set out at the beginning of the process. There is not space here to review this in detail but suffice to say that all aspects of the design were reviewed and the outcome was a more detailed design, including a breakdown of components in terms of size and weight (Figure 16.14).

On the basis of the preliminary design phase, detailed information on the performance characteristics of the aircraft was derived and in turn the data produced could be used to compare its operating costs with those of the Twin Otter and Skylander (Table 16.5). An analysis of the production cost of the aircraft using a number of methods (namely Roskam and Raymer) produced an indicative development cost of around $114 million. Depending on production runs the total cost of producing the aircraft was estimated at $1 billion for a run of 225 aircraft. Adding the development cost to the production estimate, and assuming a sales price of $4 million (similar to the expected price for the Twin Otter 400 and RUAG 228NG) for the aircraft, break-even would occur after 275 aircraft have been built.

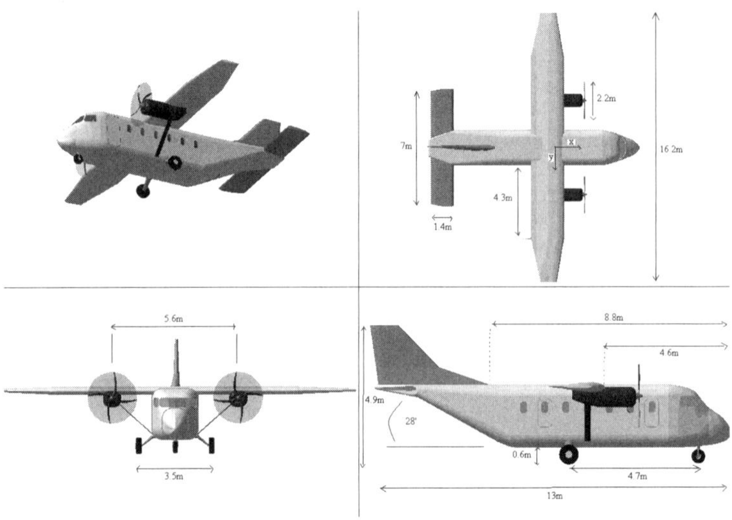

Figure 16.14 Views of the ALX-190

Source: Quéméneur (2008).

Table 16.5 Comparison data of proposed aircraft

Aircraft	MTOW	OEW	Engines
Twin Otter 400	5670 kg	3121 kg	2 × PT6A-34
Do228NG	6400 kg	3808 kg	2 × TPE331-10GP-511D
GECI Skylander	8392 kg	4664 kg	2 × PT6A-65
ALX-190	5600 kg	3125 kg	2 × PT6A-34

Sources: Quéméneur (2008); manufacturers' data.

Summary

In mid-2009 it is clear that there is a large market for a new aircraft suitable for remote regions operations to replace the considerable number of ageing aircraft in service. While there are a few types in the production phase, it is apparent that not even these aircraft will match the demand and hence it may be necessary for a new aircraft to be launched. The process described above for the ALX-190 will have to be followed and it may be that public funds will be needed to facilitate the design and production of this new design, possibly to be produced in the developing world to allow the vital services such aircraft provide to continue into the future.

References

Airliners.net (2009), http://www.airliners.net (accessed July 2009).

Flight International (2008), *World Airliner Census*, 19–25 August.

Howe, D. (2000), *Aircraft Conceptual Design Synthesis*, Wiley-Blackwell.

Quéméneur, A. (2008), *Conceptual Design of a Light Multirole Commuter Aircraft ALX-190 Lion Cub*. Unpublished MSc thesis, Cranfield University.

Raymer, D. (2006), *Aircraft Design: A Conceptual Approach*, 4th edition, AIAA.

Roskam, J. (1990), *Airplane Design part VIII: Airplane Cost, Development, Manufacturing and Development*, The University of Kansas.

Stinton, D. (2001), *The Design of the Aeroplane*, Blackwell Science.

Chapter 17
Effects of Reducing the High Price of Air Travel in Remoter Regions

George Williams and Svein Bråthen

Airfares should cover the costs of running the airports, ATM services and costs of aircraft operations. From theory (and as discussed in Chapter 1), the social optimal fare level is where the passengers pay the marginal cost of use. The air transport industry has fixed costs connected to airports, communication and aircraft operations. Even leased aircraft means fixed costs for the airlines during some period of time, and even at the departure level there are fixed costs that cannot be avoided once the aircraft is set into operation (e.g. depreciation, crew, charges for airport services and maintenance). However, the marginal costs per passenger are low (in-cabin service, extra fuel). Hence, the air transport industry is encumbered with economies of scale. Under such circumstances, marginal cost pricing would incur financial losses, because the fixed costs are not covered. Under these circumstances, a fare level equal to average costs would give total cost coverage, but a number of passengers will not have the willingness to pay a fare level significantly above marginal costs. On routes with thin traffic, fares equal to average costs would mean a fare level significantly above the marginal costs. This is depicted in Figure 17.1. Here, demand D1 shows the demand on a thin route, where the difference between the efficient marginal cost price (PM) and the average cost price P1 is large.

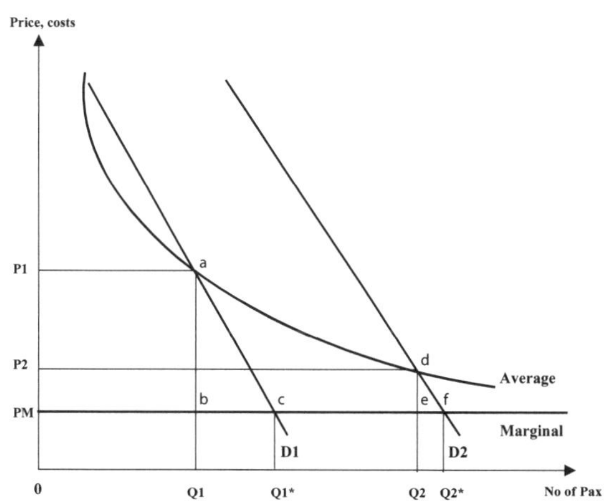

Figure 17.1 Average and marginal cost pricing, depending on market size

In such cases, there may be good reasons for introducing supporting schemes such as public service obligation (PSO) or route development funds. In this way, some of the fixed costs are paid by others and the fares can be held at such a level that the traffic volume is not too adversely affected. If demand is approaching the capacity limit of the aircraft (demand D2), then the difference between PM and the average cost price P2 is less, and PSO is likely to have a weaker justification.

In Chapter 1, pricing under financial constraints were discussed. If the revenue side is constrained in the sense that a maximum subsidy level is granted, e.g. through a PSO contract, the pricing scheme should be designed in a way that economic welfare is maximised. In the common situation where the subsidies are not high enough to cover the difference between average and marginal costs, fares have to be set above marginal costs. Welfare maximisation is reached when the departure from marginal costs pricing is done in a way that the market segment with the lowest price elasticity gets the highest fares. This is simply to say that the market with the highest willingness to pay should cover a larger proportion of the total costs. Certain fringes are used to segment the market. In practice, we find this pricing behaviour throughout the aviation industry where, for example, business travellers are paying higher fares on average. Criteria like flexible tickets, time of travel and in-cabin service are normally used as segmentation criteria.

In this chapter, we will consider the kind of effects that can be expected from price changes on subsidised routes. Two cases will be discussed, namely the experiences from Norway and Scotland. We will focus on both the effects in the traffic market, and also overall economic welfare effects.

The Effects of Price Changes in Norway

The number of PSO route legs in Norway is around 110. The number of PSO passengers was around 1.1 million in 2006 and the annual subsidy level in the period 2006–2009 was 320 million NOK. In 2006, the Ministry of Transport and Communications commissioned a study to examine the effects of a fare reduction on the Norwegian PSO routes. This fare reduction could be achieved by increasing either the PSO airline or the PSO airport subsidy level – or both.

The effects of the fare reduction are measured by means of a transport network model. This comprehensive model describes the whole transportation network, including roads, railways and sea transport. The transport costs for each transport mode is specified in detail for each route and link, also including elements like the user's own time costs, fares and vehicle operating costs. The model then predicts how the users of the whole system react to the fare reduction. These reactions may include an increased number of air trips, a change of transport mode, a change of route or even a change of destination.

The output of the analysis gives the network effects from the airfare reductions for the following elements:

- change in consumer surplus for all transport users;
- change in producer surplus for all transport companies;
- change in public sector income, spending and cost of public funding;
- external (environmental) effects.

Together, these numbers add up to show the total socio-economic effects of the airfare reduction. In addition, the geographical distribution of user benefits can be illustrated. This is most often of political interest when discussing public spending on transportation in remoter areas.

The analysis was done in two stages. Firstly, the effects of a fare reduction were analysed. Secondly, because the fare reduction causes changes to air transport demand on remoter routes, a scenario was analysed where the number of aircraft seats were adjusted in combination with the fare adjustment, to ensure a fair adaptation between capacity and demand. In the model, the number of seats was increased on the PSO routes. On some other air routes, the number of seats was slightly reduced. Here, smaller aircraft were used in the calculations but there were no changes in the departure frequencies. However, the overall picture was that competition between PSO routes and other modes of transport was a minor issue. Table 17.1 shows the results of the final analysis.

The table shows that the PSO fare reduction would entail increased subsidies of approximately 135 million NOK. This would correspond to an increase of around 40 per cent as compared to the 2006 subsidy level. On the other hand, user benefits and to some extent capacity adjustments (such as some smaller aircraft on the larger routes) will give a net positive socio-economic contribution of some 100 million NOK per year.

What can be learned from this is that even if the subsidy level may get a significant increase to compensate the airlines for revenue losses, the overall positive economic effects seem to justify this. Furthermore, the economic effects of the fare reduction are estimated to be most beneficial for people living in remoter areas like western and northern Norway. Figure 17.2 gives a simple illustration of

Table 17.1 Socio-economic effects of reduced airfares in the Norwegian PSO network – annual change, NOK (2006)

Element	**Million NOK/year (2006)**
Consumer surplus (mainly benefits from fare reductions)	241
Producer surplus, airlines and other transport companies	13
Government revenues (= increased subsidies)	-135
Costs of increased public funding (subsidies)	-25
External effects (emissions to air)	12
SUM socio-economic effects	**106**

Source: Jansson (2007).

the geographical distribution of user benefits from PSO fare reduction for affected municipalities.

The largest circles symbolise travel costs savings equivalent to the value of approximately one hour's travel time saving per trip. It can be seen that the reduced travel costs mainly devolves on the remoter regions in Northern Norway. The main objective for the PSO regime is to contribute to improved regional balance, and the research done by Jansson (2007) indicates that this objective can be further achieved by reducing the PSO fares.[1]

Figure 17.2 Geographical distribution of user benefits

1 Jansson, K. (2007). Increased subsidies on Norwegian PSO routes: Economic effects [Konsekvenser av store andel statlig finansiering av luftfarten i Norge]. Molde Research Institute/Molde University College, report 0714.

The Impact of Airfare Reductions in the Scottish Highlands and Islands

A number of initiatives have been instigated over the past decade to reduce airfares on routes serving the Scottish Highlands and Islands. The network is currently comprised of a mix of subsidised PSO and commercially viable routes (see Chapter 2), on which the latter residents of the region are able to obtain a 40 per cent fare discount. Three of the PSO routes connect Glasgow with Campbeltown, situated on the Scottish mainland on the Mull of Kintyre, and the islands of Barra and Tiree. All of the other PSO operations consist of inter-island routes within the Outer Hebrides and the Orkney and Shetland islands.

Attempts have been made to increase the number of routes in the Highlands and Islands designated as PSO services, but these have invariably failed to persuade the UK Government that such status is warranted. The most notable attempt was made in 2003 by the Highlands and Islands Transport Partnership (HITRANS) in Scotland, the statutory body comprised of regional authorities and transport providers which is responsible for pursuing improvements to the strategic transport services and infrastructure network in the north of Scotland. In a study commissioned by HITRANS the likely impact of converting all of the existing commercial operations serving the Highlands and Islands from within Scotland into PSO routes was assessed.[2] Based on an assumption of a price elasticity of demand of -0.85 for fare reductions of one-third on the Highlands and Islands network as a whole, excluding inter-island services, the study estimated that passenger traffic would increase by 28 per cent, from 383,000 in 2001 to 490,000.

To cope with this large increase in demand the study proposed a major revision to the route network and a large expansion in the number of services operated, which would have the effect of raising passenger demand by a further 45 per cent, to 719,000. The study calculated that as a result of these proposed changes the total annual cost of operating the air transport network in the Highlands and Islands would increase from £25 million to £46 million, resulting in an overall network loss of £13.4 million in 2005. Longer term, losses were predicted to be in the region of £9–11 million per annum.

The first publicised hard evidence of the effects of reducing airfares in the Highlands and Islands however came as a result of a decision of the then Scottish Executive to reduce fares in July 2004 on the three Glasgow PSO routes. Fares between Glasgow and Barra were reduced by 30 per cent, between Glasgow and Tiree by 22 per cent and by just 7 per cent between Glasgow and Campbeltown. A study undertaken by Reference Economic Consultants between September and December 2005 into the impact of the fare reductions estimated that on the Barra route passenger traffic had increased by 18 per cent and on the Tiree route by 15 per cent. By comparison the increase on the Campbeltown operation was only 3 per cent, reflecting the much smaller fare reduction. The implied price elasticities

2 An Expanded Air Services Air Network for the Highlands and Islands, Aviation & Travel Consultancy Ltd, 2003.

on the three routes were in the case of Barra -0.60, Campbeltown -0.42 and Tiree -0.69. All three routes reflected substantially more price inelastic demand than had been assumed in the Aviation & Travel Consultancy Limited study commissioned by HITRANS. In this work, price elasticity had been assumed to be -1.07 on the Barra route, -0.96 on the Campbeltown route and -1.06 on the Tiree route. Figure 17.3 shows the changes in the passenger traffic carried each year on each of the three Glasgow PSO routes between 2000 and 2008.

Air Discount Scheme

As a result of the UK Government's unwillingness to expand the PSO network in Scotland, an alternative mechanism to reduce airfares was launched by the Scottish Government in May 2006. The Air Discount Scheme (ADS) was introduced under the European Commission's Aid of a Social Character mechanism with the aim of delivering lower airfares for residents of Scotland's most peripheral communities and facilitating accessibility and social inclusion by providing them with a 40 per cent fare discount.[3] The scheme was initially approved by the Commission to run until March 2008, with a maximum budget of £11.2 million per annum.[4] The ADS has since been approved for a further three years to March 2011 with an estimated overall budget of £21.1 million.[5]

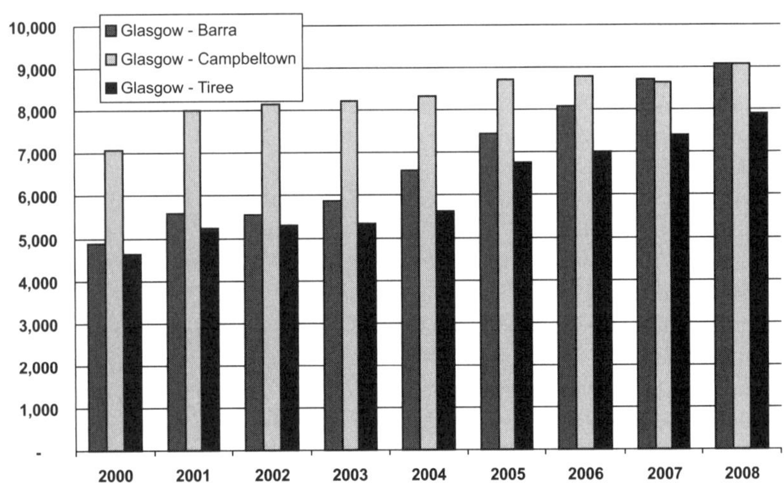

Figure 17.3 Annual passenger traffic on the Glasgow PSO routes
Source: UK CAA.

3 Up to 50 per cent discount is permitted under Aid of a Social Character.
4 State Aid No N 169/2006 (2006/1855/EC).
5 State Aid No N 27/2008 (2008/685/EC).

In order to assess the impact of ADS the Scottish Government commissioned a study by the Halcrow Group in August 2007.[6] The scheme is eligible to residents of the island communities of the Western Isles, Orkney, Shetland, Islay and Jura, and to those living on the Scottish mainland in Caithness and North-West Sutherland. Of the 100,000 people eligible to participate in the ADS, the study found that by August 2007 some 87 per cent of this group had signed up to the scheme and that one-third of the scheme's membership had purchased 130,000 discounted tickets during the first year of the scheme. Passenger traffic overall on the ADS routes increased by 12 per cent during the first year the scheme was in operation. Table 17.2 shows the populations resident in each of the main geographic areas in which the scheme is available and the associated take-up rate by those eligible to participate. Figure 17.4 shows the routes covered by the ADS.

Table 17.2 Take-up rate of the Air Discount Scheme by those eligible to participate

	Population	% participation
Western Isles	26,500	93
Orkney Isles	19,000	95
Shetland Isles	22,000	96
Islay and Jura	3,700	98
Caithness and NW Sutherland	29,100	51
Total	*100,300*	87

Source: State Aid No. N 27/2008 (2008/685/EC).

The scheme's take-up rate can best be illustrated by examining its effect on a representative selection of individual routes. Table 17.3 shows the numbers of passengers carried on the routes subject to the ADS between 2001 and 2008. As may be seen, the three busiest routes in 2008 were Aberdeen–Sumburgh (Shetland), Glasgow–Stornoway (Lewis) and Aberdeen–Kirkwall (Orkney). Passenger traffic has experienced the strongest growth between 2001 and 2008 on the following routes: Inverness–Kirkwall (157 per cent), Edinburgh–Stornoway (156 per cent), Glasgow–Kirkwall (140 per cent) and Edinburgh–Sumburgh (94 per cent). Compared to 2005, overall passenger traffic on the ADS routes by 2008 had increased by 26 per cent. The only route to have experienced a decline in passenger traffic over this period is Aberdeen–Sumburgh. Part of the reason for this fall in traffic relates to the fact that there is a substantial oil industry charter service operation linking Aberdeen with Scatsta, an airfield located close to the oil terminal at Sullem Voe on Shetland. Changes in economic conditions and in the air service supply of course have also had an influence.

6 Review of the Air Discount Scheme, Scottish Government Social Research, 2008.

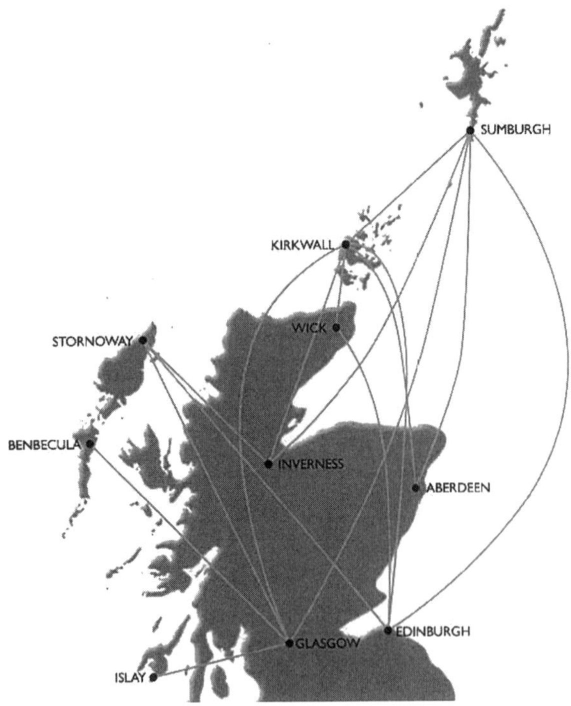

Figure 17.4 Routes included in the Air Discount Scheme

The representative routes analysed in more detail here for the effects of the ADS are Aberdeen–Kirkwall, Aberdeen–Sumburgh, Edinburgh–Stornoway, Glasgow–Stornoway and Glasgow–Islay. Figure 17.5 shows the monthly total seat capacity provided by the sole operator, Loganair, between January 2006 and September 2007 on the routes connecting Kirkwall with Aberdeen, Edinburgh, Glasgow and Inverness. Over the five summer months (May to September) of 2007 compared to the same period in 2006 total seat capacity increased by 10 per cent. Passenger traffic carried on the four routes before and after the introduction of the ADS is depicted in Figure 17.6. It is clear that the number of passengers carried had increased on all four routes as result of the ADS, the largest percentage increase being evident between Kirkwall and Glasgow. At the same time as the ADS was introduced, however, significant changes to the network of services operated from Kirkwall by Loganair were made in anticipation of increased demand. In particular, non-stop services were introduced to both Edinburgh and Glasgow. Previously, services had operated via either Inverness or Wick.

Table 17.3 Passenger traffic on the Air Discount Scheme routes

	2001	2002	2003	2004	2005	2006	2007	2008
Aberdeen–Shetland	74,046	71,079	62,341	60,845	63,004	64,680	68,817	68,357
Glasgow–Stornoway	47,932	47,823	45,918	42,034	45,853	50,238	53,790	58,144
Aberdeen–Kirkwall	36,979	40,287	40,465	38,770	39,119	42,228	42,738	44,177
Edinburgh–Kirkwall	28,853	27,977	20,446	20,697	20,423	23,226	29,229	35,656
Inverness–Stornoway	23,818	26,221	28,375	31,010	33,025	32,385	36,403	35,636
Edinburgh–Shetland	15,914	16,663	15,186	15,802	21,901	23,297	26,749	30,801
Glasgow–Islay	19,609	20,833	21,512	21,495	21,819	25,993	28,443	29,338
Edinburgh–Stornoway	11,065	13,285	24,267	31,032	29,755	27,136	25,249	28,320
Glasgow–Benbecula	21,296	19,642	20,498	18,787	20,986	22,582	24,286	25,201
Inverness–Kirkwall	9,764	14,189	15,963	16,394	18,506	22,014	25,867	25,085
Glasgow–Shetland	16,539	12,327	15,435	14,379	14,989	15,104	16,489	17,227
Glasgow–Kirkwall	6,266	5,363	5,750	6,698	6,940	11,649	15,466	15,055
Kirkwall–Shetland	7,243	7,252	6,618	5,799	6,230	7,045	8,998	8,709
Edinburgh–Wick	103	22	5,831	5,980	4,895	4,761	5,976	7,255
Aberdeen–Stornoway	–	–	-	–	–	5,305	6,750	7,113
Total	**319,427**	**322,963**	**328,605**	**329,722**	**347,445**	**377,643**	**415,250**	**436,074**

Source: UK CAA.

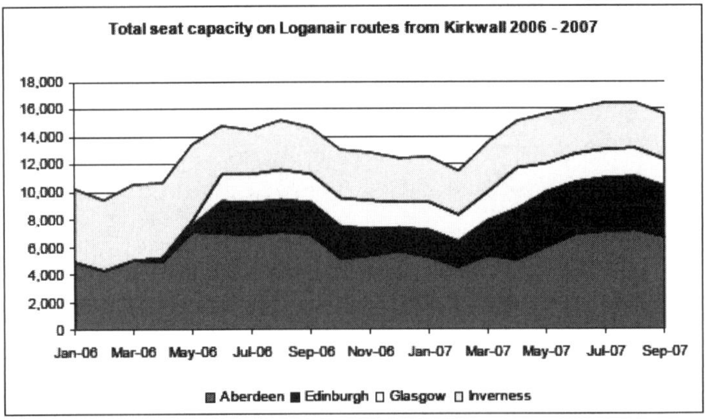

Figure 17.5 Total seat capacity on routes from Kirkwall

Source: Review of the Air Discount Scheme, Scottish Government Social Research, 2008.

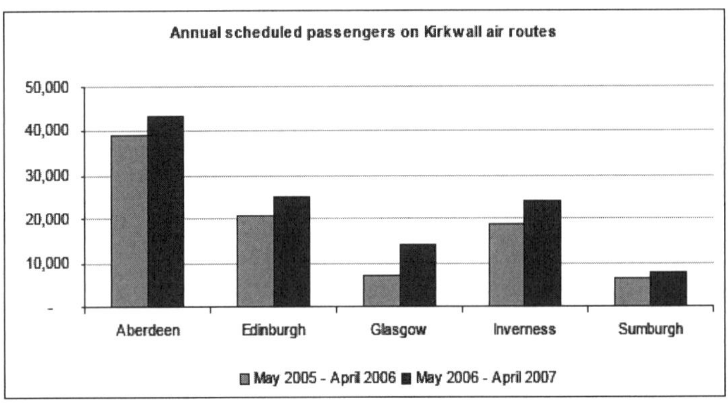

Figure 17.6 Passenger traffic on Kirkwall routes

Source: Review of the Air Discount Scheme, Scottish Government Social Research, 2008.

As may be seen in Figure 17.7, the percentage of travellers taking advantage of the discounted fares available under the ADS on the Aberdeen–Kirkwall route was 21 per cent, much lower than between Sumburgh and Kirkwall, where the equivalent figure was 57 per cent.[7]

7 Review of the Air Discount Scheme, Scottish Government Social Research, 2008, p. 27.

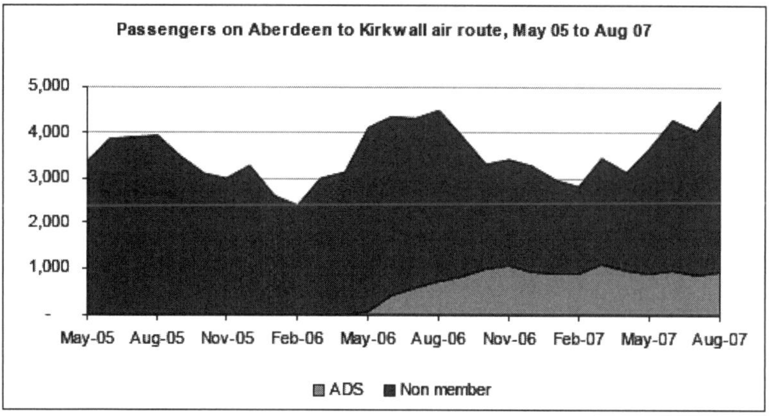

Figure 17.7 Aberdeen–Kirkwall ADS and non-ADS passenger traffic
Source: Review of the Air Discount Scheme, Scottish Government Social Research, 2008.

Figure 17.8 shows the monthly total seat capacity provided by the sole operator, Loganair, between January 2006 and September 2007 on the routes connecting Sumburgh (Shetland) with Aberdeen, Edinburgh, Glasgow and Kirkwall. Over the five summer months (May to September) of 2007 compared to the same period in 2006 total seat capacity increased by 13 per cent. Passenger traffic carried on these four routes before and after the introduction of ADS is depicted in Figure 17.9. It is clear that the number of passengers carried had increased on only two of these routes as result of the ADS, namely to Edinburgh and Kirkwall. Compared to the Kirkwall routes, on which demand increased by 25 per cent in the year following the introduction of the ADS, traffic increased on the Sumburgh routes by only 3 per cent over the same period.[8] This was despite significant changes also being made to the network of services operated from Sumburgh by Loganair. In particular, a second non-stop service was introduced to Edinburgh.

As may be seen in Figure 17.10, the percentage of travellers taking advantage of the discounted fares available under the ADS on the Aberdeen–Sumburgh route was 26 per cent, while between Sumburgh and Glasgow the equivalent figure was 39 per cent.

Figure 17.11 shows the monthly total seat capacity provided by Loganair between January 2006 and September 2007 on the routes connecting Stornoway with Edinburgh, Glasgow and Inverness. Over the five summer months (May to September) of 2007 compared to the same period in 2006, total seat capacity increased by 6 per cent. Passenger traffic carried on these three routes before and after the introduction of the ADS is depicted in Figure 17.12. It is clear that

8 Review of the Air Discount Scheme, Scottish Government Social Research, 2008, p. 29.

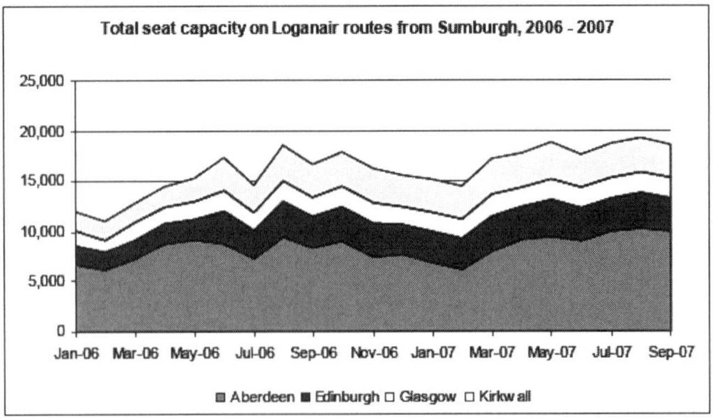

Figure 17.8 Total seat capacity on routes from Sumburgh (Shetland)

Source: Review of the Air Discount Scheme, Scottish Government Social Research, 2008.

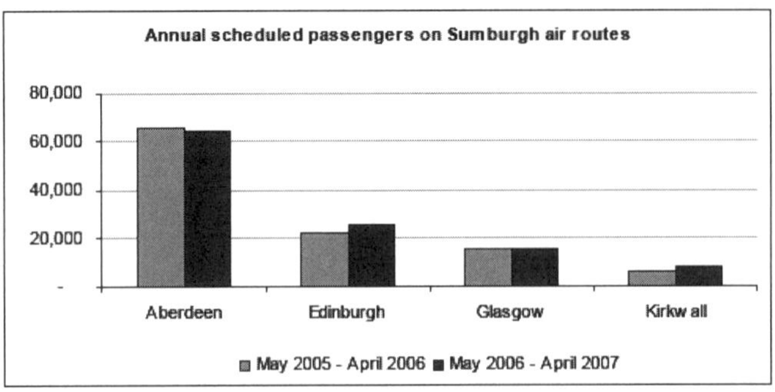

Figure 17.9 Passenger traffic on Sumburgh (Shetland) routes

Source: Review of the Air Discount Scheme, Scottish Government Social Research, 2008.

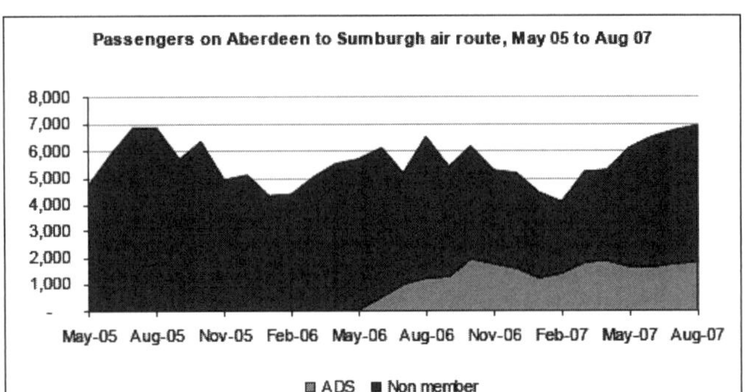

Figure 17.10 Aberdeen–Sumburgh ADS and non-ADS passenger traffic

Source: Review of the Air Discount Scheme, Scottish Government Social Research, 2008.

the number of passengers carried as result of the ADS had increased on only the Glasgow route.[9] As may be seen in Figure 17.13, the percentage of travellers taking advantage of the discounted fares available under the ADS on the Edinburgh–Stornoway route was 30 per cent, while on the Glasgow–Stornoway route the take-up rate was much higher on a route that experiences much greater seasonal variation in demand (Figure 17.14).

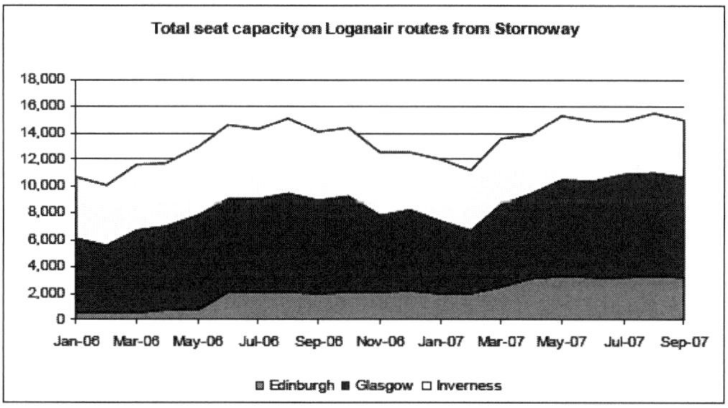

Figure 17.11 Total seat capacity on routes from Stornoway

Source: Review of the Air Discount Scheme, Scottish Government Social Research, 2008.

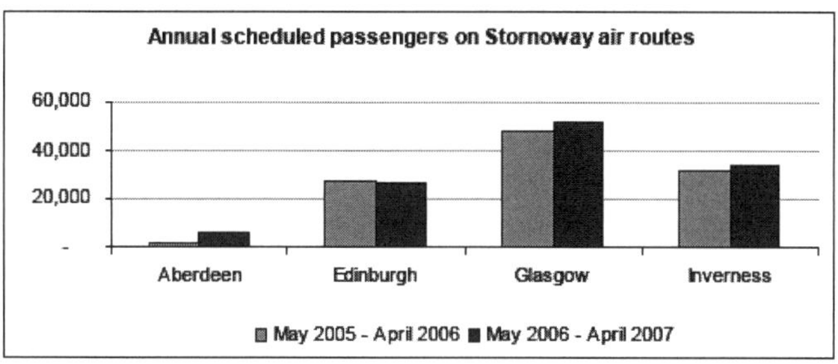

Figure 17.12 Passenger traffic on Stornoway routes

Source: Review of the Air Discount Scheme, Scottish Government Social Research, 2008.

9 The largest increase in demand was evident between Stornoway and Aberdeen, the route not having been operated before 2006.

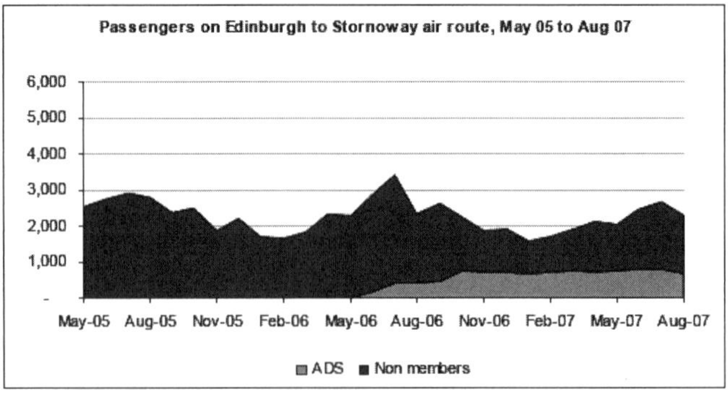

Figure 17.13 Edinburgh–Stornoway ADS and non-ADS passenger traffic
Source: Review of the Air Discount Scheme, Scottish Government Social Research, 2008.

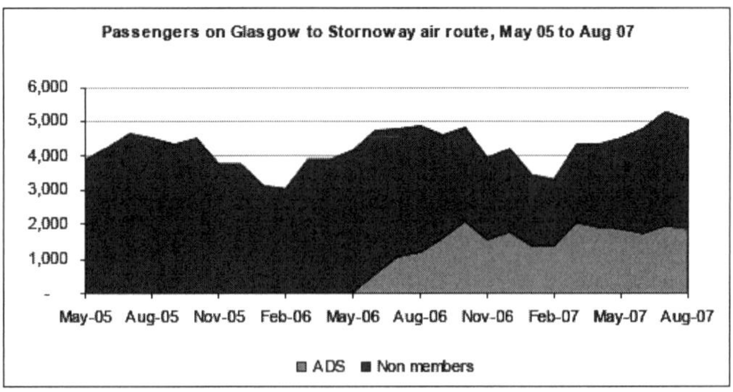

Figure 17.14 Glasgow–Stornoway ADS and non-ADS passenger traffic
Source: Review of the Air Discount Scheme, Scottish Government Social Research, 2008.

In the first four months of 2007 compared to the same period in 2006 passenger traffic between Glasgow and Islay increased by 25 per cent, while the total seat capacity provided by Loganair increased by 15 per cent. Figure 17.15 shows the percentage of travellers taking advantage of the discounted fares available under the ADS on the route.

Some 440,000 discounted tickets have been purchased since the ADS was introduced in May 2006.[10] Due to the fact that there were significant changes to the route network and the number of services operated by Loganair both shortly before and after the introduction of the ADS, it is not possible to produce any accurate

10 The Scottish Government's Air Discount Scheme, Martin Ritchie, 6th Forum on Air Transport in Remoter Regions, Bergen, April 2009.

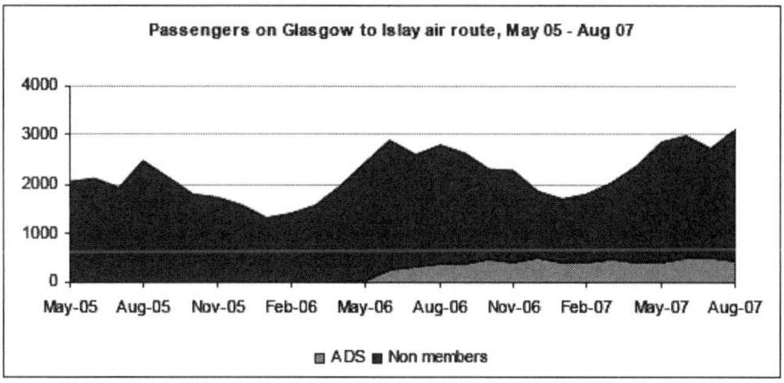

Figure 17.15 Glasgow–Islay ADS and non-ADS passenger traffic

Source: Review of the Air Discount Scheme, Scottish Government Social Research, 2008.

estimate of fare elasticity. On the basis of the evidence available following the implementation of the ADS, however, it would seem more likely that the elasticity levels revealed when fares on the Glasgow PSO routes were reduced are much more likely to reflect reality than the estimates produced by Aviation & Travel Consultancy Limited for HITRANS in 2003. A full socio-economic assessment of the ADS will be carried out for the Scottish Government in 2010, with a view to determining future policy with regards to the provision of affordable air services for the communities living in the Scottish Highlands and Islands.

Index